the science *of* self-empowerment

ALSO BY GREGG BRADEN

BOOKS

Resilience from the Heart
The Turning Point
Deep Truth
The Divine Matrix
Fractal Time
The God Code
*The Isaiah Effect**
Secrets of the Lost Mode of Prayer
The Spontaneous Healing of Belief

AUDIO PROGRAMS

An Ancient Magical Prayer (with Deepak Chopra)
The Turning Point
Awakening the Power of a Modern God
Deep Truth
The Divine Matrix
The Divine Name (with Jonathan Goldman)
Fractal Time
*The Gregg Braden Audio Collection**
Speaking the Lost Language of God
The Spontaneous Healing of Belief
Unleashing the Power of the God Code

DVDS

The Science of Miracles

*Available from Hay House except
items marked with an asterisk.
Please visit:

Hay House USA: www.hayhouse.com®
Hay House Australia: www.hayhouse.com.au
Hay House UK: www.hayhouse.co.uk
Hay House India: www.hayhouse.co.in

the science of self-empowerment

awakening the new human story

GREG BRADEN

HAY HOUSE, INC.
Carlsbad, California • New York City
London • Sydney • New Delhi

Published in the United States by: Hay House, Inc.: www.hayhouse.com®
Published in Australia by: Hay House Australia Pty. Ltd.: www.hayhouse.com.au
Published in the United Kingdom by: Hay House UK, Ltd.: www.hayhouse.co.uk
Published in India by: Hay House Publishers India: www.hayhouse.co.in

Cover design: Charles McStravick • *Interior design:* Pamela Homan
Indexer: Shapiro Indexing Services

Grateful acknowledgment is made to the Institute of HeartMath for permission to reproduce the steps for "Attitude Breathing," copyright © 2013 Institute of HeartMath, and the steps for "Quick Coherence," copyright © Institute of HeartMath.

Additional grateful acknowledgment is made for illustrations licensed through Dreamstime stock images, member of PACA and CEPIC.

Library of Congress has cataloged the earlier edition as follows:

Names: Braden, Gregg, author.
Title: Human by Design: from evolution by chance to transformation by choice / Gregg Braden.
Description: Carlsbad, California : Hay House, Inc., 2017. | Includes bibliographical references.
Identifiers: LCCN 2017013536 | ISBN 9781401949310 (hardback)
Subjects: LCSH: Human beings--Philosophy. | Human beings--Origin. | Spirituality. | BISAC: BODY, MIND & SPIRIT / New Thought. | BODY, MIND & SPIRIT / Inspiration & Personal Growth. | SCIENCE / Life Sciences / Evolution.
Classification: LCC BD450 .B6423 2017 | DDC 128/.6--dc23 LC record available at https://lccn.loc.gov/2017013536

Tradepaper ISBN: 978-1-4019-4932-7
E-book ISBN: 978-1-4019-5725-4
Audiobook ISBN: 978-1-4019-5584-7

10 9 8 7 6 5 4 3 2 1
1st edition, October 2017
2nd edition, March 2019

Printed in the United States of America

SFI label applies to the text stock

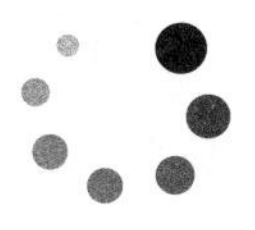

"For small creatures such as we, the vastness is bearable only through love."

—Carl Sagan (1934–1996),
American astronomer and cosmologist

CONTENTS

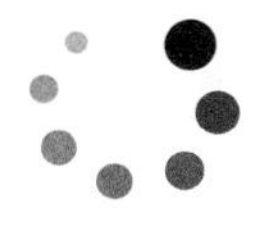

AUTHOR'S NOTE

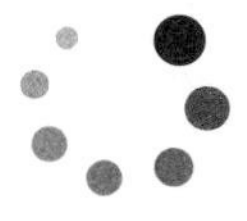

In Part II of this book, I use the term *wired* to indicate that we already have the biology we need and that we are predisposed to accomplish the extraordinary potentials described in each chapter.

Wired is a slang term that also has had other meanings in the past.

The original usage can be traced back to the days before telephones, when the telegraph was the primary mode of communication. At that time it was common to say that we had "wired" a message to someone, meaning that we had sent a telegraph message. Later meanings have varied from the jitters caused by ingesting too much caffeine or certain drugs to the way the neurons in our brains are connected. It's for these reasons that I'm clarifying my intent for the word right up front, before it's used in the following pages.

INTRODUCTION

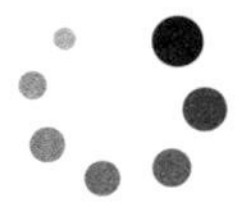

Our Origin: Why It Matters

Since our earliest ancestors looked with awe into the distant stars of a moonless night sky, a single question has been asked countless times by countless numbers of people sharing the same experience through the ages. The question they've asked speaks directly to the core of every challenge that will ever test us in life, no matter how big or how small. It's at the heart of every choice we'll ever face, and it forms the foundation for every decision we'll ever make. The question that's at the root of all questions asked during the estimated 200,000 years or so that we've been on earth is simply this: *Who are we?*

In what may be the greatest irony of our lives, following 5,000 years of recorded history and technological achievements that stagger the imagination, we have yet to answer this most basic question with certainty.

Key 1: In the presence of the greatest technological advancements of the modern world, science still cannot answer the most fundamental question of our existence: *Who are we?*

WHY OUR ORIGIN MATTERS

The way we answer the question of how we came to be as we are penetrates the essence of each moment of our lives. It forms the perceptual eyes—*the filters*—through which we see other people, the world around us, and most importantly, ourselves. For example, when we think of ourselves as separate from our bodies, we approach the healing process feeling like powerless victims of an experience that we have no control over. Conversely, recent discoveries confirm that when we approach life *knowing* that our bodies are designed to constantly repair, rejuvenate, and heal, this shift in perspective creates the chemistry in our cells that mirrors our belief.[1]

Our self-esteem, our self-worth, our sense of confidence, our well-being, and our safety stem directly from the way we think of ourselves in the world. From the person we say yes to when it comes to choosing a life partner and how long our relationship lasts once we create it, to what jobs we feel worthy of performing, the most important decisions we'll ever make in life are based in the way we answer the simple, timeless question *Who are we?*

On a more spiritual level, our answer creates the foundation for the way we view our relationship with God. It even justifies our thinking when it comes to trying to save a human life, and when we choose to end one.

The way we think of ourselves is reflected in what we teach our children as well. When their delicate sense of self-worth is threatened by relentless bullying from rivals and classmates, for example, it's their answer to *Who am I?* that gives them the strength to heal their hurt. Their answer can even make the difference between when they feel worthy of living and when they don't.

On a larger scale, the way we think of ourselves determines the policies of corporations and nations that either justify dumping 12-million-plus tons of used plastic and thousands of gallons of radioactive waste into the world's oceans each year or show we cherish the living oceans enough to invest in preserving them.

Even how countries choose to create the borders that separate them and how our governments justify sending armies across those borders onto the land and into the homes of the people of another nation begins with the way we view ourselves as people. When we think about it, our answer to the most basic question—*Who are we?*—is at the core of everything we do and defines all that we cherish.

Key 2: Everything from our self-esteem to our self-worth, our sense of confidence, our well-being, and our sense of safety, as well as the way we see the world and other people, stems from our answer to the question *Who are we?*

It's precisely because the way we think of ourselves plays such a vital role in our lives that we owe it to ourselves to explain who we are and where we came from as truthfully and honestly as possible. This includes taking into consideration every source of information available to us, from the leading-edge science of today to the wisdom of 5,000 years of human experience. *This also includes changing the existing story when new discoveries give us the reasons to do so.*

WHY WE NEED A NEW STORY

Over 150 years ago, geologist Charles Darwin published a paradigm-shattering book entitled *On the Origin of Species by Means of Natural Selection,* often shortened simply to *Origin of Species.* His book was intended to provide a scientific explanation for the complexity of life—how it has morphed over the ages from primitive cells to the complex forms we see today. Darwin believed that the evolution he witnessed in some parts of the world, in some forms of life, applied to all life everywhere, including human life.

In one of the great ironies of the modern world, since Darwin's time the very science that was expected to support his theory, and eventually solve life's mysteries, has done just the opposite. The most recent discoveries are revealing facts that fly in the face of long-standing scientific tradition, especially when it comes to human evolution. Among these facts are the following.

Fact 1: The relationships shown on the conventional human evolutionary tree—the dashed lines that connect one fossil to another and lead to modern humans at the top of the tree—are not based on evidence. While these relationships are believed to exist, they have never been proven and are *inferred* or *speculative* relationships.

Fact 2: Modern humans arose suddenly on earth approximately 200,000 years ago with the advanced features that set us apart from all other known forms of life *already developed.*

Fact 3: The lack of common DNA between ancient Neanderthals, thought to be some of our ancestors, and early humans, whose DNA is similar to ours, tells us that we did not originally descend from the Neanderthals, even if we interbred with them at some point.

Fact 4: Advanced genome analysis reveals that the DNA that sets us apart from other primates is the result of an ancient, mysterious, and precise fusion of genes that suggests something *beyond* evolution made our humanness possible.

To be clear, the advanced features identified in Fact 2 didn't develop slowly over long periods of time, as evolutionary theory would suggest. Instead, characteristics that include a brain 50 percent larger than that of our nearest primate relative and a complex nervous system with emotional and sensory abilities fine-tuned to our world already existed in modern humans when they appeared. And humans haven't changed.

In other words, contemporary humans are the same humans, 2,000 centuries later!

These facts, which are based upon peer-reviewed science, present a problem for the long-held evolutionary story of our origins. The new evidence clearly doesn't support the conventional narrative of the past that we've been taught. The popular story that's being shared in our classrooms and textbooks today leads us to believe we're insignificant beings that began as a biological fluke long ago, then endured 200,000 years of brutal competition and "survival of the strongest," only to discover that we're powerless victims in a hostile world of separation, competition, and conflict.

The scientific discoveries described in this book, however, now suggest something radically different. It's for this reason that we need a new story to accommodate the new evidence. Or conversely, we need to follow the evidence we already have to the new story that it tells.

Before his death in 1962, Nobel Prize–winning physicist Niels Bohr reminded us that the key to solving a mystery is found within the mystery itself. "Every great and deep difficulty bears in itself its own solution," he said. "It forces us to think differently in order to find it."[2] Bohr's words are just as powerful today as when he spoke them over half a century ago.

From fossils and grave sites to brain size and DNA, the existing evidence is already solving the mystery of the origin of our kind. It's already telling us our new story. The key is that we must first think differently about ourselves in order to accept what the story reveals. I've written this book as an invitation to do just that.

Key 3: By allowing new discoveries to lead to the new stories they tell, rather than forcing them into a predetermined framework of ideas, we may, at last, answer the most important questions of our existence.

WHY THIS BOOK?

The purpose of this book is to 1) reveal new discoveries about our origin, in Part I, and 2) show how to apply these discoveries in our day-to-day lives, in Part II. Rather than speculate upon how the first cell of life appeared upon the earth, I'll begin as Darwin did, *at the time following our mysterious origin.* Both Part I and Part II include exercises to help anchor the significance of specific discoveries in your own life.

WHAT THIS BOOK IS NOT

- *Science of Self-Empowerment* is not a science book. Although I will share the leading-edge science that invites us to rethink our relationship to the world, this work has not been written to conform to the format or standards of a classroom science textbook or a technical journal.
- *Science of Self-Empowerment* is not a religious book. It's not intended to support any particular religious belief regarding creation or human origins, such as Creationism. *Science of Self-Empowerment* is based upon the peer-reviewed scientific evidence (anthropological, paleontological, biological, and genetic) that begins *immediately after* our species appeared on earth. As such, there are places where the new story in this book may appear contrary to the traditional stories of religion, as well as to those of traditional science.
- *Science of Self-Empowerment* is not a peer-reviewed research paper. Each chapter *has not* gone through the lengthy review process of a certified board or a selected panel of experts conditioned to see our world through the eyes of a single field of study, such as physics, math, or psychology.

WHAT THIS BOOK IS

- This book *is* well researched and well documented. I've written *Science of Self-Empowerment* in a reader-friendly way that incorporates true-life accounts, scientific discoveries, and personal experiences to support an empowering way of seeing ourselves in the world.
- This book *is* an example of what can be accomplished when we cross the traditional boundaries between science and spirituality. By marrying state-of-the-art discoveries of biology, genetics, and earth sciences with ancient wisdom, we gain a powerful framework for understanding what's possible in our lives.

NEW DISCOVERIES MEAN A NEW STORY

If we're honest with ourselves and acknowledge that the world is changing, then it makes sense that our story in the world must change as well. In all likelihood, the new human story will be a hybrid of theories that already exist. These will be woven together into the new tapestry of a grand chronicle that describes an extraordinary and epic past. And with this new story, at last we'll embrace the history that cannot be accounted for in any of the existing theories by themselves.

A growing body of evidence suggests that we are the product of something more than random mutations and lucky biology. But the evidence can only go so far. Fossils, DNA, ancient cave art, and human burial sites can only show us the remnants of what happened in the past. They cannot tell us why those things happened. Unless we find a way to travel backward in time, the truth is that we may never know the complete *why* of what has made our existence possible.

But maybe we don't need to know. Maybe it's not necessary to have that level of detail for us to shift the way we think about ourselves and change our lives. The discovery that we are the product of something more than evolution—very likely a conscious and intelligent act of creation—may be all we need to point us in a new, honest, and healthy direction when it comes to the human story.

The undeniable fact is that something happened 200,000 years ago to make our existence possible. And whatever that something was, it left us with the extraordinary abilities of intuition, compassion, empathy, love, self-healing, and more.

We owe it to ourselves to embrace the body of evidence, the story it tells, and the healing it can bring to our lives. The power of the emerging human story may help us bring true and lasting healing to the racial hate, the sexual violence, the religious intolerance, and the other devastating challenges we face, ranging from the abuse of technology to the plague of terrorism that's sweeping the earth. To do anything less is simply to place a Band-Aid on the emotional wound that creates these expressions of fear.

For the first time in the 300-year-long history of science, we're writing a new human story that gives us a new answer to the timeless question of who we are.

Key 4: New DNA evidence suggests that we're the result of an intentional act of creation that has imbued us with extraordinary abilities of intuition, compassion, empathy, love, and self-healing.

This book is written with one purpose in mind: to empower us in the choices that lead to thriving lives in a transformed world.

Gregg Braden
Santa Fe, New Mexico

PART I

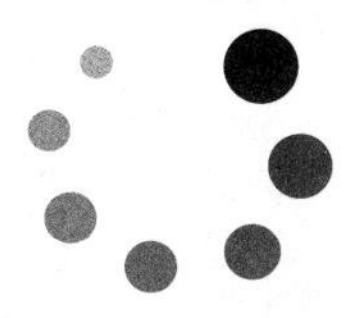

the new human story

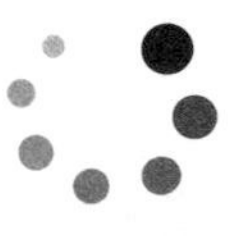

The purpose of the following chapters is to empower you with new ways of thinking and new reasons to think differently about yourself and the relationships in your life: the relationships that you have with other people, the relationship that you have with the earth and the world around you, the relationship that you have with yourself, and ultimately, the relationship you have with God / Spirit / Universal Source / the One. Before you discover these empowering implications, however, it helps first to establish what you believe right now—a baseline for the way you think about yourself and your place in the world.

The following exercise is not intended to judge you or criticize any existing thinking, feelings, or beliefs. It's simply a point of reference to identify beliefs that you may not be aware of, or to clarify beliefs that you may have only suspected in the past.

EXERCISE

Establishing a Baseline for Your Beliefs

Using your answers to the following questions as a place to begin, by the end of the book you will easily see where and how the new information you've learned has transformed the way you think of yourself and your potential. For this exercise, you'll need paper and pen.

The Technique. Using single words or brief phrases, please write down your answers to the following questions as honestly as you can. For yes-or-no questions, circle your answer.

- **Questions about Your Origins.**

    1. Do you believe that the origin of life, in general, is the result of a chance event that happened long ago, as conventional science suggests?

        Yes No

2. Do you believe that human life in particular is the result of a chance event that happened long ago, as evolution theory suggests?
   Yes No

- **Questions about Your Potential.**

3. Do you believe that you're designed to consciously influence the events of your life, the quality of your life, and how long you live?
   Yes No

   If you answered no, continue to "Defining Your Beliefs" below.

   If you answered yes, please answer questions 4 through 6:

4. Do you trust your ability to trigger self-healing in your body on demand when you need it?
   Yes No

5. Do you trust your ability to trigger your deepest states of intuition when you need them, on demand?
   Yes No

6. Do you trust your ability to self-regulate your immune system, your longevity hormones, and your overall health?
   Yes No

- **Defining Your Beliefs.** Finish the following sentences.

7. When I notice something unusual happening with my body (sudden aches or pains, an unexplained rash, a rapid heartbeat for no apparent reason, and so forth) I find myself *feeling* ____________________________.

8. When I notice something out of the ordinary happening with my body, *the first thing I will do is* ____________________________.

chapter one

Breaking Darwin's Spell

Evolution Is a Fact—Just Not for Humans

"Who are we . . . but the stories we tell about ourselves, particularly if we accept them?"

— Scott Turow (1949–), American author

"Why are you here?" a voice asked from somewhere in the darkness.

From what sounded like a faraway place, a man was asking the question, yet he sounded so distant that I wasn't sure if he was speaking to me or to someone else. I remember the sensation of feeling both awake and asleep at the same time and thinking that maybe I was dreaming. It didn't even occur to me that I could open my eyes to see who the man was. Then I heard his voice again, this time speaking my name. "Gregg . . . you're okay. You did great. But I need you to tell me why you're here." This time I knew I wasn't dreaming—the man knew my name and he was speaking directly to me. Instinctively, my eyes began to open as I turned my head in his direction. The light overhead was so bright that it forced me to squint as I looked up at the ceiling from my bed. Surprisingly, the man wasn't far away at all. In fact he was standing right next to me, looking down at me from behind a blue surgical mask. Seeing him jolted my memory and suddenly I remembered what was happening.

I was awakening from the anesthesia I'd been given earlier that morning. I was in the post-operation recovery room of the Mayo Clinic in Jacksonville, Florida. The voice I was hearing was the doctor who had reassured me only an hour or so earlier that I was in good hands with his team, and that I would be okay. And while he continued his assurances, I wasn't prepared for the question he kept asking about why I was there.

Less than a month before, an examination at a different clinic had shown an anomalous growth on the wall of my bladder. "Something is in your bladder that shouldn't be there," that first doctor had told me. "It needs to be removed." Wanting to ensure the best possible outcome for whatever was needed, I'd gone to the prestigious Mayo Clinic for a second opinion. It was there that I discovered that the only way to determine with certainty that the growth was benign was to test the tissue itself—to perform a biopsy.

What was happening now, however, was not part of the original plan. After being fully anesthetized and prepped for surgery, I was waking up to a puzzled doctor asking a question that I could barely answer in my altered state of consciousness: Why was I there? He was asking the question because the anomalous growth that had shown up in the previous exams was no longer there. The surgeon was telling me there was nothing to remove because I had a normal and healthy-looking bladder. To emphasize his point, he showed me a color photograph of the inside of my bladder, taken only moments before.

As I did my best to grasp what he was saying, the surgeon used the tip of his pen as a pointer to show me where the growth had been in the earlier scans. He emphasized, however, that today there was no bruising, no discoloration, and no scar tissue or any sign whatsoever to indicate that anything out of the ordinary had ever existed. And he wanted to know why. He wanted to know how such a thing could have happened.

In my groggy state I was not as eloquent with my answer as I would have liked. I did my best to tell the doctor about the research I'd done into the self-healing potential of the human body, the ancient traditions that had mastered this healing potential, and the science that now confirms that our bodies can heal themselves when given the conditions to do so. The last memory I have of this doctor is of him turning away and walking toward the door as I tried my best to answer his question. The explanation I was offering for what we'd both experienced that day was obviously not what he had expected, nor what he wanted, to hear.

When I thought about my doctor's response later, after my recovery, I could understand his frustration. There is absolutely nothing in the training of a modern medical professional that allows for us to have such self-healing relationships with our bodies. And it's for precisely this reason that when an experience such as mine occurs, the medical team has limited options when it comes to offering an explanation. They generally chalk it up to a mistaken diagnosis, an unexplainable spontaneous recovery, or simply a miracle.

From my doctor's point of view, a miracle had just happened in his operating room and he was trying to make sense of it. From my point of view, however, what had happened was less about a miracle and more about a technology—a powerful inner technology that's available to each of us—whose existence has been largely forgotten over time.

Since 1986, I have researched the wisdom, studied the principles, and where possible, experienced the techniques embraced by ancient and indigenous traditions when it comes to our ability to self-heal. From the monks, nuns, and abbots in the monasteries of Tibet, Nepal, and Egypt to the indigenous healers and shamans of the Yucatán jungles in Mexico and the Andes Mountains of southern Peru, our ancient ancestors, and their modern counterparts, have done their best to preserve the knowledge of the most intimate relationship we can ever have: our relationship with our own bodies. And while the knowledge they preserved is not science in the traditional sense, new scientific discoveries in genetics, molecular biology, and the new fields of epigenetics and neurocardiology have confirmed many of the relationships described in the ancient traditions.

When it came to my own body, however, even though I strongly believed that self-healing was possible and had even witnessed other people's success at it, a combination of my scientific training and the limiting beliefs instilled in me at an early age by my alcoholic father and dysfunctional family environment had left a deep doubt that such a healing was possible for me. So even though I'd performed the yogic techniques, qigong, and other healing modalities, taken the medicinal herbs, adopted a raw diet, and accepted emotional changes to the best of my ability between my diagnosis and the procedure at the Mayo Clinic, I still doubted my capacity to create for myself the successful healings that I'd seen occur with others. And it was because of my doubt that I had chosen the modern technology offered through one of the highest-rated medical facilities in the world as a responsible option for the diagnosis I'd received.

As a trained scientist, I cannot say to you that the practices, techniques, and lifestyle changes I adopted during those two weeks were the reason the medical team found nothing to remove the day of my surgery. What I can say is that new scientific discoveries have identified a link between specific healing modalities known in the past and their ability to restore balance in our bodies. It's the fact of this relationship that invites an honest reassessment of the limiting story we've been told about our origin as a species and what we're capable of. When we consider the facts revealed by the best science of today, spontaneous healings and miracles such as the one I experienced seem less rare and extraordinary and more like an ordinary part of everyday life. The chapters that follow reveal these discoveries and the story they tell. And with that larger story, we're given the reasons to embrace a new answer to the question *Who are we?* and to write our new human story.

If you've ever felt that there's more to the story of our past than we've been led to believe, I want you to know you're not alone. A 2014 Gallup poll revealed that in the United States alone, a whopping 42 percent of the people who were asked believe that there's something more to human origins than is typically acknowledged in the mainstream—*that something beyond Charles Darwin's theory of evolution is responsible for our existence.*[1] The results of this poll reflect a growing sense that we humans are part of something great, powerful, and mysterious. Some of the greatest minds in science agree.

SOMETHING IS MISSING FROM THE HUMAN STORY

Francis Crick, the Nobel Prize–winning co-discoverer of the DNA double helix, believed that the eloquence of life's building blocks has to be the result of something more than a lucky quirk of nature. Through his pioneering research, he was one of the first humans to witness the complexity and the sheer beauty of the molecule that makes life possible. Late in life, Crick risked his reputation as a scientist by publicly stating, "An *honest* man, armed with all the knowledge available to us now, could only state that in some sense, the origin of life appears at the moment to be almost a miracle."[2] In the scientific world, this statement is the equivalent of heresy, suggesting that something more than chance evolution led to our existence.

The feeling that there's something more to our story is not just a recent phenomenon. Archaeological discoveries show that, almost universally, ancient humans felt connected to more than just their immediate surroundings. They sensed that we have our roots in other worlds, some that we can't even see, and that we are ultimately part of a cosmic family that lives in those worlds.

The sacred text of the ancient Mayan *Popol Vuh*, for example, describes how the "Forefathers" created humankind, while the Christian Bible and the Hebrew Torah describe how we are the descendants of wise and powerful beings linked to a greater and otherworldly intelligence.[3,4,5] Could there be a simple explanation as to why such a sense has remained with us so strongly, across such diverse traditions, and has lasted for so long? Is it possible that our feeling of having an intentional origin and a greater potential is based in something that's true?

When we ask *Who are we?* the short answer is that we're not what we've been told and we're more than most of us have ever imagined.

WE ARE A SPECIES OF STORIES

From the time of our earliest ancestors, we've used stories to explain the world around us and describe our place in it. Sometimes our stories are based in fact. Sometimes they're not. Some stories are metaphorical. We've used these stories to explain the unexplained and make sense of our existence.

The ancient Egyptians, for example, thought of the earth, the space beneath the earth, and the sky above as worlds unto themselves. In their view of creation, the earth beneath their feet was floating upon Nun, a primordial ocean that was the source of the Nile River. The sky above was formed by the body of the goddess Nut. The dome of Nut's rounded belly was the home of the sun and the stars as she arched over the earth, facedown, throughout time. The realm under the earth, Duat, was where the sun would go at night as it disappeared beneath the horizon at sunset.[6]

All these realms had deities—gods and goddesses—associated with them that played a powerful role in the daily lives of the Egyptian people. And while the stories weren't based in science, they worked for the people of the time. They provided a mechanism to explain what the ancient Egyptians saw happening in their everyday world and helped them know where they fit in.

Today, we continue to use stories to explain our world. And our stories play a role that's more important than ever. Not only do they inform the way we manage everything from disease and healing to our relationships and romances; on a global level, the future of our planet and the survival of our species, which now hang in the balance, also depend on the stories we choose to embrace. It's precisely for these reasons that it's vital we tell ourselves the right story.

OUR STORIES DEFINE OUR LIVES

We cherish the stories we create. As individuals we often proudly share our family history and the accomplishments of our ancestors. As nations we defend with pride our teams' athletic achievements at the Olympics, the scientific and technical advancements that sent our astronauts to the moon, and the flags that unite us as countries. But sometimes we find ourselves defending stories that we've grown up with even when new discoveries tell us these stories are wrong. It's our willingness to cling to a story that's familiar, even if new evidence shows us it's obsolete, that may be the greatest hurdle we face as we learn to embrace our world of extremes in a healthy way.

Key 5: The stories that we tell ourselves about ourselves—and believe—define our lives.

A commonly used axiom suggests that if we hear something said often enough we begin to accept that something as fact, whether or not it's true. The sanitized story of smoking tobacco that was generally accepted until the early 1960s is a perfect example. Prior to a 1964 report on the dangerous effects of cigarette smoking, America's tobacco companies were engaged in a powerful media campaign to convince the public that smoking was a safe, even a healthy, habit. Catchy slogans such as "When tempted to over-indulge, reach for a Lucky instead," "I protect my voice with Luckys," and "As your dentist I would recommend Viceroys" were common in magazine, radio, and television advertisements.[7]

A particularly disturbing poster for Camel cigarettes from the 1940s stated that, according to a nationwide survey, "More doctors smoke Camels than any other cigarette."[8] A further investigation into the survey revealed the rest of the story. The questions had been asked of doctors who had received complimentary packs of Camel cigarettes at meetings and conferences before they took the survey. It was after they'd received the free samples that they were asked what brand they liked best or had in their pockets. The samples effectively skewed the answer in favor of Camels. American consumers trusted and believed these and other ads. After all, if a cigarette was safe for doctors, it must be safe for everyone else, right?

The perception of such messages, and of tobacco use itself, however, changed forever with the landmark study from the surgeon general. For the first time, the study reported scientifically what many people had suspected intuitively. It described a direct link between tobacco use, chronic bronchitis, and lung cancer. The study stated, "It is the judgment of the committee that cigarette smoking contributes substantially to mortality from certain specific diseases and to the overall death rate."[9] By 1965, the tobacco industry was required to place the now-familiar warning labels on every tobacco product sold.

The point of this example is to illustrate that a belief once shared by the mainstream media and the general public—the story that smoking tobacco is safe—changed over time. It had to change because the evidence of debilitating diseases experienced by so many tobacco users simply didn't fit the mainstream story of safety and health. It didn't jibe with what people actually experienced.

WE'RE SOLVING 21ST-CENTURY PROBLEMS WITH 19TH-CENTURY THINKING

In a similar way, an information campaign to skew public opinion is happening today when it comes to us and the story of our origin. The 19th-century theory of human evolution is taught as undisputed fact in today's classrooms, leaving no room for consideration of any other possible explanation for the mystery of our existence. And because the mainstream story does not take into account recent discoveries, it leaves us unprepared to address the radical social issues and global challenges we're experiencing today, including everything from terrorism, bullying, and hate crimes to the epidemic of drug and alcohol abuse among young people.

Because we are invested in the theory of evolution, we use it to guide our decisions, and so we celebrate competition and force over cooperation and compassion. Among other things, we keep trying to solve problems associated with our racial, religious, and sexual diversity through the obsolete thinking of competition and "survival of the strongest"—both of which are key components of the theory of evolution. It makes no sense when we think about it, and yet, for reasons of habit, money, ego, and power, the mainstream educational system and educators cling to an outdated story of human origins that's no longer supported by the evidence. Both the tobacco story and the story of human origins illustrate perfectly why it's important to get our stories right—and what can happen when we don't.

CHANGE THE STORY, CHANGE YOUR LIFE

When it comes to the human family, the shared stories of our successes, the memories of our tragedies, and the inspiring examples of our heroism are the threads that connect us. Our connection is powerful, primal, and necessary. Whether it's the big issues of politics, religion, or shipping weapons to "freedom fighters" in war-torn countries half a world away, or deeply personal issues such as the right of a gay man to marry or a woman's right to control her own body, modern technology now allows us to share the stories that justify our choices and the future we want to create.

English novelist Terence David John Pratchett, known to his fans as Terry Pratchett, beautifully described the awesome power of our stories when he said, "Change the story, change the world."[10] I think there's a lot of truth in this statement. Our lives are reflections of what we believe about ourselves and how the world works. Pratchett's observation is so universal, in fact, that we can take it one step further.

In the same breath that we say, "Change the story, change the world," we can go to an even deeper level by saying, "Change the story, *change our lives*." Both statements are true. And both offer a powerful way of thinking in the darkest moments of our lives.

Key 6: When we change the story, we change our lives.

The scientific narrative regarding the vastness of the cosmos, and our insignificance in it, is a perfect example of the powerful influence that a story can have on us. It also illustrates the axiom that if we tell a story enough times we begin to accept it as true.

THE OLD STORY: SMALL, POWERLESS, AND INSIGNIFICANT

For the last century and a half we've been steeped in a cosmic story that leaves us feeling like little more than trivial specks of dust in the universe, or biological sidebars in the overall scheme of life. Carl Sagan described this mind-set perfectly when he commented on the scientific perspective on our place in the cosmos: "We find that we live on an insignificant planet of a humdrum star lost in a galaxy tucked away in some forgotten corner of a universe in which there are far more galaxies than people."[11]

This kind of limited thinking, promoted by the scientific community, has led us to believe that we're unimportant when it comes to life in general and also separate from the world, from one another, and ultimately, even from ourselves.

Albert Einstein echoed this perception of our insignificance when it came to his ideas about the validity of the evidence in the emerging field of quantum physics that suggested that all things are deeply connected. Einstein couldn't accept the fact of that connection. Leaving no doubt in our mind as to what he believed the new quantum ideas meant for science, Einstein said, "If quantum theory is correct, it signifies the end of physics as a science."[12] His beliefs wouldn't allow him to actually accept the possibility that we live in a world where everything and everyone is so intimately linked.

One of the reasons for Einstein's resistance to the ideas of the new physics was that to live in a world of quantum connection would mean we have the ability to influence what happens in our lives and are faced with the responsibility for the outcomes we create. Ultimately, it was Einstein's firm belief that we live in a world where things are not connected that prevented him from fulfilling his life's dream. He passionately believed that his research would eventually lead him to discover a scientific truth that united all the laws of nature, a "theory of everything." Sadly, Albert Einstein died in 1955 without seeing his elusive dream realized.

With Einstein and Sagan's legacies of separation and human insignificance in mind, it's not surprising that we often feel helpless when it comes to what happens in our bodies and lives. In a world of disconnection, we're told that things just happen whenever and however they do. Is it any wonder that we often feel powerless when we see the world changing so fast that some say it's "falling apart at the seams"?

Charles Darwin's proposal regarding human evolution in the mid-1800s laid the foundation for the scientific conclusions of our insignificance that came later, in the early 1900s. The theory of evolution was based upon the premise that we are the latest result of a series of chance events that have never been witnessed, proven, or duplicated, and we can attribute the fact that we still exist to the "survival of the strongest" among us. The theory that struggle has gotten us to where we are today suggests that we're hopelessly locked into lives of competition and conflict. Culturally, this idea is now accepted to such a degree that many people believe that using force is the best way to do things in the workplace and in the community of nations.

Consciously, and sometimes on levels that are unconscious, this belief of struggle and conflict plays out every day in our lives. And it happens sometimes in surprising, unexpected ways. For example, when we find our "hot buttons" being triggered by those who know us best in our most intimate relationships, even the most spiritually minded among us will lash out, using hurtful tactics to protect ourselves in the moment. The reason is not surprising.

From the time we're born, and even before, while we're still in our mothers' wombs, we begin to learn how to cope with the world through the thoughts and feelings of our caregivers. We learn from the tone of our mothers' voices, for example, when the world is safe and when it's not. We also learn to associate the chemicals of stress, as well as the chemicals of pleasure, that flow through our bodies with the voices, sounds, and experiences that trigger the release of those chemicals.

Unless we are fortunate enough to come from a really healthy family of caregivers, the chances are good that their responses to the world are based upon the false conditioning they learned from the caregivers in their early lives. And it's precisely these patterns from other people, sometimes generations old, that become our patterns as well.

So when we feel threatened as adults, it's these conditioned patterns that show up in whatever way our minds deem necessary for our survival. When the patterns kick in, they draw from the deep well of whatever beliefs are "hardwired" into our subconscious minds. The key here is that these beliefs are often rooted in the stories and experiences of other people.

Do we lash out violently, as we're conditioned to do through our stories of "survival of the strongest"? Or do we respond confidently and honestly, embracing the deeper knowledge of our connection with all life, including our connection with the people who've just triggered us?

To be absolutely clear, I'm not suggesting that either response is right or wrong, good or bad. I am saying, however, that our reactions don't lie. Regardless of what we may think we believe, the way we respond in such intimate moments is a telling reflection of what we truly believe. The point here is that the stories we're told during our most vulnerable and impressionable years of childhood form our most deeply held beliefs. And that's where the story of our origins comes in too.

A TALE OF TWO ORIGINS

We begin hearing the story of human origins early in life. And depending upon our families' beliefs, sometimes we're even exposed to two entirely different and conflicting stories taught around the same time—one at home and the other at school.

In most schools we're taught the scientific theory of evolution by natural selection, which is a sterile and unsettling story for any young person to hear. It begins long ago with an unbelievable run of good luck, when *just* the right atoms combined at *just* the right time to create *just* the right molecules under *just* the right conditions to lead to the first simple forms of life that would eventually become the complexity of us.

Even the most passionate supporter of evolution must admit that the uncannily good fortune required for such a series of events requires a stretch of imagination, or faith, that such a process is even possible. As noted previously, Francis Crick called the existence of DNA "almost a miracle."

Evolution theory accounts for this good fortune, however, suggesting that it's the struggle itself—the competition among varying forms of life—that made this unlikely combination of events successful. Proponents of evolution claim that competition has led us to be the present-day winners in nature's multimillion-year-long quest to survive. The key here is that we're told that "struggle" has served us well in the past, and by extension, still serves us today. In fact, struggle has been so successful, we're told, that it's actually been "programmed" into our bodies genetically. So because of natural selection, we're now supposedly hardwired for competition and struggle.

At the same time that children are learning the scientific story of evolution and struggle in school, they're often told a religious story that's equally frightening. This story also begins at the time of our beginning. And it also requires a stretch of the imagination to believe that it is even possible. In Judaism, Christianity, and Islam, this story is the story of a mysterious force—God—and how God created the first human from the dust of the earth, breathed life into the being he created, and caused the first human, Adam, to wake upon Earth.

From this story we learn that we are the descendants of Adam and his children, and that we come into this world inherently flawed as people. The rest of the story describes how we're destined to struggle between good and evil as we search for a way to redeem ourselves from our flaws. Other world religions use similar stories to explain the origin of humankind and the purpose of life.

Both stories—the scientific and the religious—begin long ago. Both have mysterious gaps in the details. And both leave us feeling separate from the rest of our world. Perhaps most importantly, both stories leave us with the feeling that we exist as we do on earth today as unwitting combatants locked into a hopeless struggle for survival—either with nature or between good and evil. From either the scientific or the religious point of view, as different as the stories may seem on the surface, when we look a little deeper, we realize that they start from the same place and have the same purpose. They begin with the fact that we exist as we do, and they are attempts to explain what our ancient existence means for us today.

Despite emerging evidence that does not fit with the traditional scientific story, educators perpetuate the theory of evolution and human survival, and teach it in our classrooms, as if it were an absolute and undisputed fact. And this is where the problem begins: We're trying to solve modern problems that require cooperation and mutual aid through a 150-year-old story based in competition and struggle. Not surprisingly, the story we've embraced—the theory of evolution—no longer makes sense in addressing where we come from and how we've become as we are. We need a new human story that reflects the new evidence in order to break the spell Darwin's ideas have on us.

BREAKING DARWIN'S SPELL

Darwin published *Origin of Species*, his best-known book, in 1859. From the time of its publication until today, the implications of this book have reverberated through the foundation of our society. Whether it's the academic controversy of where we come from and why we're here, or the emotionally charged issues of conception, abortion, and the death penalty that sometimes polarize families and whole communities, the implications of Darwin's work impact our lives in a way few other ideas can. I often wonder if Darwin himself ever imagined the effect that his work would have on the world and how deeply his ideas would influence the lives of everyday people living over a century into his future.

Before *Origin of Species,* there were few sources to turn to when it came to answering life's biggest questions. Prior to the mid-19th century, the philosophical questions of life, such as *Where do we come from? Why are we here?* and *How do we make life better?* were relegated to religion and traditional folklore. With the publication of Darwin's first book, this changed. The theory of evolution offered a new story to answer life's big questions that didn't require biblical interpretations or religious teachings.

Key 7: For the first time in recorded human history, Charles Darwin's theory of evolution, published in 1859, allowed science to answer the big questions of life and our origin without the need for religion.

While the full title of Darwin's book, *On the Origin of Species by Means of Natural Selection,* may sound complex, the idea that it's based upon is really very simple. Darwin proposed that all life, including human life, began with a single primal organism that mysteriously appeared on earth long ago. Darwin didn't even attempt to describe how that organism first came into existence. In fact, contrary to what many people commonly assume, the actual origin of life was never his focus. While he readily acknowledged that the science of his day had yet to shed any meaningful light on that mystery, he also admitted that solving the mystery of how life began wasn't necessary for his theory of evolution to be accepted.

Darwin defended his beliefs by using the analogy of another unsolved mystery to make his point. He pointed to the scientific acceptance of gravity as an analogy for how it's possible to accept a theory even though it hasn't been fully explained. "It is no valid objection," he said, "that science as yet throws no light on the far higher problem of the essence or origin of life. Who can explain what is the essence of the attraction of gravity? No one now objects to following out the results consequent on this unknown element of attraction."[13]

From this and similar statements, it's clear that Darwin was less concerned with *how* life originally appeared and more concerned with *what happened after* it did so. Specifically, how did the simple form of life that he believed first emerged in the world morph into the complexity and diversity that we see as life today?

Darwin based his theory of evolution on his personal experience and direct observations. Many of those observations were made during a five-year journey aboard the British research ship the HMS *Beagle*.[14] Darwin was the designated naturalist on the ship, whose mission sounds much like the mission to document new forms of life in unknown galaxies for the starship *Enterprise* (of *Star Trek* fame). His job was to document new forms of life in the uncharted lands discovered during the *Beagle*'s voyage. Although Darwin's journey lasted from 1831 to 1836, he didn't share his theory until 23 years later. With the publication of *Origin of Species*, for the first time the essence of Darwin's theory of evolution was available to the general public. He writes:

> But if variations useful to any organic being do occur, assuredly individuals thus characterized will have the best chance of being preserved in the struggle for life; and from the strong principle of inheritance they will tend to produce offspring similarly characterized. This principle of preservation, I have called, for the sake of brevity, Natural Selection.[15]

Today, over 150 years after Charles Darwin first published his theory, the best scientists of the modern world, from the best universities of our time, having access to the most funding in research history and using the most advanced technology ever available, are still struggling to prove the viability of this theory in general, and specifically when it comes to humans.

In essence, the unanswered questions are:

- Does evolution alone explain the diversity that we see in the natural world today?
- Does evolution apply to humans?

As we'll see in the sections that follow, new discoveries are making it necessary to rethink the way we've answered both these questions in the past.

EVEN DARWIN HAD HIS DOUBTS

Charles Darwin didn't know in his day what we know today about the world. He couldn't have. Many fields of science that we take for granted simply didn't exist until later in the 19th century and early in the 20th century. Darwin couldn't have known about genetics, for example. While the fact that one generation can inherit the traits of its parents was recognized during Darwin's time, exactly what made the transfer possible—DNA—was not understood until after his death. Darwin couldn't have known about the specialized heart cells that give us access to the extraordinary abilities and sensitivities that will be described later in the book. And he couldn't have known that those cells, or the capabilities they make possible, already existed when modern humans appeared on the scene 200,000 years ago.

While Darwin couldn't have known these things specifically, he clearly suspected that future discoveries would overturn at least some of his theory. He stated this possibility in his writings. In *Origin of Species*, he writes: "If it could be demonstrated that any complex organ existed which could not possibly have been formed by numerous successive slight modifications"—the hallmark of evolution—"my theory would absolutely break down."[16]

It's because the conditions that Darwin himself described as the keystone to his theory have now been overturned—because in fact we do have complex organs that did not form through "numerous successive slight modifications"—that evolution theory, alone, cannot explain what we find in the real world. In other words, just as Darwin suspected would happen, his theory has broken down.

In *Origin of Species* Darwin revealed his suspicion that evolution theory might not be enough to explain the complexity of life. Though the following statement may appear a bit wordy, it's Darwin's language. I'm sharing it so that you'll have a sense of his reservations—in this case, with regard to the complex functions of an eye.

> To suppose that the eye with all its inimitable contrivances for adjusting the focus to different distances, for admitting different amounts of light, and for the correction of spherical and chromatic aberration, could have been formed by natural selection, seems, I freely confess, absurd in the highest degree.[17]

The fact that the complexity of the eye, as well as the complexity of a number of other organs, meets the condition that Darwin himself stated would invalidate his theory, opens the door to the theme for Part I of this book: Evolution in and of itself is not enough to account for the extraordinary features and abilities we've had from the beginning. The evidence suggesting that certain physical features—including our eyes, our advanced nervous systems, and our brains—were already functional when modern humans arose is casting doubt on Darwin's theory when it comes to humankind.

HUMAN EVOLUTION: SPECULATION TAUGHT AS FACT

The conventional thinking of today leaves us with the sense that Darwin's theory of evolution is a "done deal." That it's an open-and-shut case universally accepted by the scientific community and there is little room for doubt when it comes to the explanation of life as we see it today. Evolution is described as fact in textbooks and classrooms. In this environment of unconditional acceptance, scientific discoveries that cast doubt on evolution are often not reported, or worse yet, are ridiculed as superstition, religion, or pseudoscience. For this reason, people are often surprised when there is any mention of discoveries casting doubt on Darwin's theory.

A perfect example of this one-sided view is the choice by the Public Broadcasting Service (PBS) to exclude any competing scientific theories or scientific criticism of evolution in their beautifully produced eight-hour miniseries *Evolution: A Journey into Where We're from and Where We're Going,* which aired in 2001. In the network's own words, the goals of the program were to "heighten public understanding of evolution and how it works, to dispel common misunderstandings about the process, and to illuminate why it is relevant to all of us."[18] And for anyone watching the series, they did just that, illustrating evolution solely from Darwin's perspective, which many scientists see as flawed for reasons that will be described later in this chapter.

A review of the PBS special by author and former White House speechwriter Joshua Gilder minced no words with regard to the way the content was produced: "The problem [with the PBS documentary] is that none of it is true, or is so fraught with inconsistencies, misinterpretation, and bad (sometimes fraudulent) data as to be worthless as science."[19] Gilder based his critique, in part, upon the scientific discoveries documented by molecular biologist Jonathan Wells in his book *Icons of Evolution,* where the PBS "proofs" of human evolution are dismantled one by one.

TAKING EVOLUTION TO THE COURTS

The evolution controversy is especially visible when it comes to state and national laws regarding what teachers are allowed to teach in public schools. A recent Senate bill in the state of Oklahoma is a perfect example of this. In 2016, Republican senator Josh Brecheen introduced legislation to allow teachers to encourage their students to think critically about the topics that affect their lives and their future.

Brecheen's proposed legislation, Senate Bill 1322, states that the purpose of the legislation is to "create an environment within public school districts that encourages students to explore scientific questions, learn about scientific evidence, develop critical thinking skills and respond appropriately and respectfully to differences of opinion about controversial issues. . . . Teachers shall be permitted to help students understand, analyze, critique and review in an objective manner the scientific strengths and scientific weaknesses of existing scientific theories covered in the course being taught."[20]

While Brecheen's bill does not mention the teaching of evolution specifically, it's clear from his history of introducing similar legislation since his election in 2010, and the inclusion of the phrase *scientific theories*, that his goal was to allow teachers to share discoveries related to human origins, including discoveries that don't support the existing story of evolution.

In 2005, the legal ruling informally known as the Dover Case was about evolution specifically, and about the way a new, alternative theory of human origins known as *intelligent design* relates to evolution. The case made worldwide headlines because it was the first legal test of the new theory in a U.S. federal court.

The Dover Case began when eleven families filed a lawsuit against the Dover Area School District of York County, Pennsylvania, over a change in the required curriculum for a ninth-grade biology class. In 2004, the school district had directed teachers to offer discoveries supporting intelligent design in addition to the traditional teaching of Darwin's theory of evolution. Proponents of the theory of intelligent design, which was first used in the book *Of Pandas and People* in 1989, assert that "certain features of the universe and of living things are best explained by an intelligent cause, not an undirected process such as natural selection."[21] Both theories were being offered in the classroom as possible explanations for human origins. The parents who filed the suit felt, however, that the ideas of intelligent design were too similar to the religious ideas of creationism, a belief that the universe and living organisms originate from acts of divine creation, so they demanded that the teaching of the new theory be discontinued.

The case was heard as a bench trial, rather than a trial by jury, and the outcome immediately sparked controversy when the judge ruled that the conclusions drawn from the science-based discoveries underlying intelligent design were, in fact, not science at all.

From the United States District Court for the Middle District of Pennsylvania, with John E. Jones III (appointed by George W. Bush in 2002) as the sitting judge at the time, the finding reads as follows:

> Teaching intelligent design in public school biology classes violates the Establishment Clause of the First Amendment to the Constitution of the United States (and Article I, Section 3, of the Pennsylvania State Constitution) because intelligent design is not science and "cannot uncouple itself from its creationist, and thus religious, antecedents."[22]

Immediately following the trial there were accusations of false testimony, even perjury, when it came to the details and expert witnesses called to reveal the scientific evidence for intelligent design. Due to the nature of a bench trial, where there are no jurors, the religious and political beliefs of the judge, and the questionable testimony, the controversy continues today.

To be absolutely clear, I'm not suggesting that intelligent design is the answer to the mystery of human origins or that the trial should not have happened. What I am saying is that I believe we owe it to ourselves to be honest about any new discoveries that are made and to consider where they may lead. What is troubling about this court ruling is what appears to be a double standard used to discount the science that supports intelligent design. On one hand, the 150-year-old theory of evolution—one that has yet to be scientifically proven—is taught as fact. On the other hand, scientific evidence suggesting that the theory of evolution is incomplete or leading us in the wrong direction is not even allowed to be mentioned in the classroom.

When we're denied the opportunity to question existing theories and present new ones based upon new evidence, we also lose the power of critical thinking that we will need if we are to successfully confront the challenges of today's world and survive those of the future.

It's the authoritative nature of beautiful and convincing documentaries, such as PBS's *Evolution*, and the skewed nature of the legal arguments, such as those made in the Dover trial, that lead many people to believe that Darwin's theory of evolution is an open-and-shut case for natural selection. Nothing could be further from the truth.

While many scientists have, in fact, accepted evolution as the best theory to explain the mystery of human origins, so far their acceptance does not exclude the recognition of new theories, especially when the new theories are anchored in good science.

I've included the objections to evolution in *Science of Self-Empowerment* for two reasons:

1. To give visibility to the fact that Darwin's theory of evolution is not an accomplished fact when it comes to science explaining who we are
2. To give voice to a sampling of the esteemed scientists who object to evolution theory in a way that is not reflected in the mainstream media today

In the remainder of this chapter, I'll share some of the opinions that continue to fuel the fires of controversy regarding the theory of human evolution.

ONE HUNDRED FIFTY YEARS OF OBJECTIONS

Passionate objections to Darwin's theory appeared almost as soon as his book was published in 1859. The first was raised by Louis Agassiz, who is regarded as one of the great scientists of the 19th century. His pioneering legacy is recognized in the field of natural history, specifically for his work in the areas of geology, biology, paleontology, and glaciology. His tireless dedication to his work took such a priority in his overall life that he once declared to a colleague, "I cannot afford to waste my time making money."[23] In other words, he was so consumed with his research and making discoveries about the natural world that making a living was secondary. While he and Darwin were both using the same methods and looking at the same information, their interpretations couldn't have been more different.

Commenting on Darwin's theory in an 1874 publication, Agassiz wrote, "The world has arisen in some way or another. *How it originated is the great question, and Darwin's theory, like all other attempts, to explain the origin of life, is thus far merely conjectural.* I believe he has not even made the best conjecture possible in the present state of our knowledge."[24]

Agassiz was not alone in his objections. A community of respected scientists has objected to Darwin's work from the time it was first published. That community continues to grow. Its roster now sounds like a who's who of leading minds in contemporary science. Following is a sampling of the types of criticisms that have been raised from the time Darwin introduced his theory in 1859 to the present.

"Darwin's theory is not inductive—not based on a series of acknowledged facts pointing to a general conclusion."[25]

— Adam Sedgwick (1785–1873), Cambridge University, British geologist and one of the founders of modern geology

"There are . . . absolutely no facts either in the records of geology, or in the history of the past, or in the experience of the present, that can be referred to as proving evolution, or the development of one species from another by selection of any kind whatever."[26]

— Louis Agassiz (1807–1873), Harvard University, American geologist

"The theory suffers from grave defects, which are becoming more and more apparent as time advances. It can no longer square with practical scientific knowledge, nor does it suffice for our theoretical grasp of the facts. . . . No one can demonstrate that the limits of a species have ever been passed. These are the Rubicons which evolutionists cannot cross. . . . Darwin ransacked other spheres of practical research work for ideas. . . . But his whole resulting scheme remains, to this day, foreign to scientifically established zoology, since actual changes of species by such means are still unknown."[27]

— Albert Fleischmann (1862–1942), University of Erlangen, German zoologist

"Evolution became in a sense a scientific religion; almost all scientists have accepted it and many are prepared to 'bend' their observations to fit with it."[28]

— H. S. Lipson (1910–1991), University of Manchester Institute of Science and Technology, British physicist

"Evolution is the backbone of biology and biology is thus in the peculiar position of being a science founded on unproven theory. Is it then a science or a faith? Belief in the theory of evolution is thus exactly parallel to belief in special creation. Both are concepts which the believers know to be true, but neither, up to the present, has been capable of proof."[29]

— Leonard Harrison Matthews (1901–1986), Cambridge University, British zoologist

"The chance that higher life forms might have emerged in this way is comparable with the chance that a tornado sweeping through a junkyard might assemble a Boeing 747 from the materials therein. I am at a loss to understand biologists' widespread compulsion to deny what seems to me to be obvious."[30]

— Sir Fred Hoyle (1915–2001), Cambridge University, British astronomer; formed the theory of stellar nucleosynthesis

"Ultimately the Darwinian theory of evolution is no more or less than the great cosmogenic myth of the twentieth century. The truth is that despite the prestige of evolutionary theory and the tremendous intellectual effort directed towards reducing living systems to the confines of Darwinian thought, nature refuses to be imprisoned. In the final analysis we still know very little about how new forms of life arise. The 'mystery of mysteries'—the origin of new beings on earth—is still largely as enigmatic as when Darwin set sail on the Beagle."[31]

— Michael Denton (1943–), British biochemist, senior fellow, Center for Science and Culture

"But how do you get from nothing to such an elaborate something if evolution must proceed through a long sequence of intermediate stages, each favored by natural selection? You can't fly with 2 percent of a wing or gain much protection from an iota's similarity with a potentially concealing piece of vegetation. How, in other words, can natural selection explain the incipient stages of structures that can only be used [as we now observe them] in much more elaborated form?"[32]

— Stephen Jay Gould (1941–2002), Harvard University, American paleontologist and evolutionary biologist

"The point, however, is that the doctrine of evolution has swept the world, not on the strength of its scientific merits, but precisely in its capacity as a Gnostic myth. It affirms, in effect, that living beings create themselves, which is, in essence, a metaphysical claim. . . . Thus, in the final analysis, evolutionism is in truth a metaphysical doctrine decked out in scientific garb."[33]

— Wolfgang Smith (1930–), American mathematician and physicist

The preceding statements offer insights rarely seen by the public, and certainly not shared in typical school classrooms, when it comes to accepting Darwin's theory. In 2001, during the same period of time that PBS was airing the *Evolution* miniseries, a diverse group of international scientists signed a declaration that they posted online to let the world know that, for them, the mystery of our origins was not yet solved. As of July 2015 the declaration had been signed by 1,371 esteemed scientists from around the world and the list of signatories continues to grow.

The petition itself is brief and simply reads:

> We are skeptical of claims for the ability of random mutation and natural selection to account for the complexity of life. Careful examination of the evidence for Darwinian theory should be encouraged.[34]

Clearly, the jury is still out on the viability of Darwin's theory of evolution when it comes to solving the mystery of human beginnings. It's obvious from objections such as the ones listed, and more, that criticism of evolution continues with passion and vigorous debate. And while Darwin's ideas are a century and a half old, they're still among the most emotionally charged issues of our time. My sense is that the reason for the controversy is twofold: first, the theory has deep moral, social, and religious implications; second, evolution is usually presented as scientific fact even though conflicting issues have yet to be resolved.

HONORING CHARLES DARWIN

Now that we have viewed some of the objections to Darwin's theory of evolution, I'd like to take this opportunity to clarify my personal view as a geologist, researcher, and author when it comes to Charles Darwin himself and his ideas of evolution.

I'll begin by stating that I have tremendous respect for Charles Darwin, both as a man and as a scientist, for what he accomplished in his day. He lived in a society that was very different from our 21st-century world. It took tremendous courage for him to offer what he did, in the way that he did it, during his time in history. The Catholic Church played a powerful and dominant role in 19th-century England, and Darwin knew that his theory would pose a direct threat to the religious doctrine of the Church. It was precisely because of this awareness that he waited over twenty years after his voyage on the HMS *Beagle* ended in 1836 to publish his book. In a letter he wrote to botanist Asa Gray in 1860, he stated his concern, saying that he "had no intention to write atheistically."[35]

Darwin lived to see his fears of such criticism justified as Cardinal Henry Edward Manning, England's highest-ranking Catholic official when *Origin of Species* was published, attacked the theory of evolution as a "brutal philosophy," stating that it implied "the ape is our Adam."[36] In spite of such criticism, at the time of his death in 1862, Darwin was considered to be the greatest scientist of his era.

I'd also like to acknowledge that much of the controversy that Darwin's theory has caused both in his time and today is due to 1) a misunderstanding of what he actually said and 2) the desire of universities, college professors, the scientific community at large, and politicians to hold his work sacred and infallible. In other words, institutions and the people who support them have attempted to make Darwin's work into something he himself never intended it to be. They want to use his theory for purposes he never foresaw or intended.

Darwin was a geologist and, by all accounts, a good geologist. He was fair and honest when he wrote about what he observed, as well as about what he believed his observations were telling him. His work was well thought out and meticulously documented, and his methods followed the accepted guidelines of the period. Where I believe Darwin's process was flawed is in regard to what he did after he published *Origin of Species*. Because his theory of evolution seemed to fit what he saw happening for one form of life in one place in the world—specifically, for the finches of the Galapagos Islands—he tried to generalize the theory to apply to all life everywhere, including humankind. This leap is where Darwin's theory of evolution appears to break down.

While we still don't know precisely what did happen when our modern human ancestors appeared 200,000 years ago, the best evidence we've obtained from the fossil record does not support evolution as the explanation for how they came to be as they were. I'm mentioning this point now because the thinking that is perpetuated by the mainstream media and many academic institutions that have a vested interest in keeping the story of evolution alive is that the controversy is over.

A THEORY IN NEED OF PROOF

Immediately following Charles Darwin's 1859 release of *Origin of Species,* the widespread acceptance of his theory led to a search for the physical evidence to support it: the "missing links" between species that were believed to exist in the fossil record. If scientists could find these clues, the logic goes, then they would be able to reconstruct our ancient family tree of development. Just the way we can document our individual family lineage in reverse, going from our parents to our grandparents, and then to our great-grandparents, and so on, they assumed one day it would be possible to create a family tree of all our collective ancestors.

The current thinking about our human evolutionary tree is shown in Figure 1.1. In this image, modern humans are represented by *Homo sapiens,* the bold dot in the upper left portion of the chart. The lines forming the branches that connect us with the other skulls lower on the tree represent the various paths of development—evolutionary paths—scientists believe have led from early primates to us today.

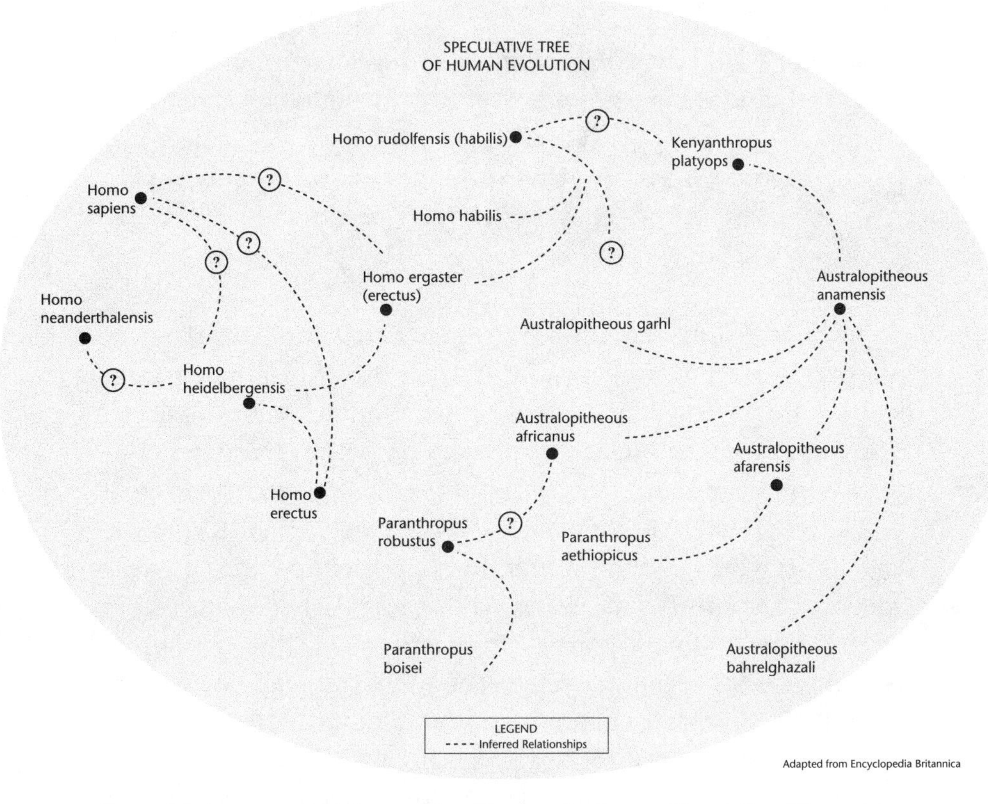

Figure 1.1. An example of the traditional human family tree of evolution. The problem with the thinking represented by this tree is that physical evidence confirming a connection between the fossils has yet to be discovered. This lack of evidence is the reason the lines that form the tree are labeled "inferred" relationships.

A close look at the illustration in Figure 1.1, however, reveals that the links between the fossils are shown as dashed lines rather than solid ones. This means that the lines represent *speculative* or *inferred* connections rather than proven ones. While the links are believed to exist, after 150 years of searching for the evidence to support them, they have yet to be proven.

Key 8: While the connections between ancient primates and modern humans on the evolutionary family tree are believed to exist, they have never been proven as fact—they are inferred and speculative connections only, at this point in time.

In other words, the physical evidence that confirms the evolutionary links that influence aspects of our lives ranging from health care to the moral justification of hate crimes, suicide, assisted suicide, and the death penalty as well as the criteria for our self-image and intimate relationships has yet to be discovered!

From the time evolution theory was introduced in 1859 to the date of this writing, to the best of this author's knowledge no clear evidence of a transitional species leading to us—that is, fossils that reflect an evolutionary journey from primitive to more human-like beings—has been discovered! Thomas H. Morgan, the winner of the 1933 Nobel Prize in physiology and medicine, left no doubt of this in the minds of readers of his book *Evolution and Adaptation*. As modern science applies what Morgan says are the "most rigid . . . tests used to distinguish wild species," he states, "Within the period of human history we do not know of a single instance of the transformation of one species into another."[37]

In the face of impassioned scientific debates, and with the "futuristic" technology that's now unlocking life's deepest mysteries, the stark fact of Morgan's observation remains a warning against wholeheartedly embracing the theory of human evolution. Even so, the theory continues to be taught in public classrooms as if it's an undisputed fact!

In *Origin of Species*, Darwin acknowledged the irony in the lack of physical evidence to support his theory. He also noted that the reason for the lack of physical evidence could possibly be explained in one of two ways: Either the geologists were interpreting the history of the earth incorrectly, or he himself had incorrectly interpreted the observations that became the foundation of his theory.

In Darwin's own words:

> Why does not every collection of fossil remains afford plain evidence of the gradation and mutation of the forms of life? We meet with no such evidence, and this is the most obvious and forcible of the many objections which may be urged against my theory.[38]

It's against the backdrop of these ideas and criticisms that an astounding discovery in the late 20th century gave scientists the opportunity to put some of the strongest-held arguments for evolution to the test. If human evolution has in fact occurred, as Darwin's theory hypothesizes, then the best way to prove the theory would be to compare us to our ancestors at the deepest level of our cells. To do so, scientists would need to sample the DNA of our early ancestors and compare it to the DNA of our bodies today, which is a problem because modern humans have already been on earth for 200,000 years. Because DNA is fragile, it doesn't last that long.

Is it possible that DNA from ancient primate life could still exist today? And if it were to exist, could we test the recovered DNA the way we routinely test our DNA today? Although these questions sound as if they could have come from the plot of *Jurassic Park*, a movie depicting ancient dinosaurs being resurrected through DNA in the present day, the answer to these questions came to light in the form of a one-of-a-kind discovery in 1987. The revelations of the discovery have left more questions unanswered, created even deeper mysteries, and opened the door to a possibility that has been forbidden territory in traditional science.

chapter two

Human by Design

The Mystery of Fused DNA

"All of us who study the origin of life find that the more we look into it, the more we feel that it is too complex to have evolved anywhere."

— Harold Urey (1893–1981), Nobel Prize–winning chemist

On Saturday, February 28, 1953, two men walked into the Eagle pub in Cambridgeshire, England, and announced a discovery that would forever change the world and the way we think of ourselves. At noon that day, Cambridge University scientists James Watson and Francis Crick announced to their colleagues who were having lunch at the pub, "We have discovered the secret of life!"[1] Watson and Crick had just made their breakthrough discovery of the double helix pattern of the DNA molecule—nature's code for life.

DNA is held within each cell of our body in threadlike structures that are called chromosomes. As humans, we have 23 pairs of chromosomes in our cells. Each chromosome, in turn, is made of smaller, more precise regions of DNA called genes. It's the codes contained within the genes and chromosomes that determine everything about the function of our bodies, including the regulation of hormones and blood chemistry, how fast and to what size our bones grow, the size of our brains, the kind of eyes we have, and how long we live—even automatic functions such as breathing, digestion, metabolism, and body temperature. With a discovery of this magnitude, it would seem that the greatest mysteries of our existence would be solved. Many have been. However, due to the deeper insights that DNA discoveries have made possible, scientists now face a quandary when it comes to interpreting where the new information about our genetic code fits into the accepted human story.

RETRIEVING DNA FROM A NEANDERTHAL BABY

In 1987, a paradigm-shattering discovery was made in the Caucasus region of Russia, near the border between Europe and Asia. Buried deep in the earth, in a place called Mezmaiskaya Cave, scientists discovered the remains of a Neanderthal infant—a baby girl that lived about 30,000 years ago! For reference, the last ice age ended about 20,000 years ago, meaning that this baby was alive during the ice age. Her remains were in an extremely rare state of preservation, and scientists were able to determine her age as somewhere between that of an unborn seven-month fetus and a two-month-old infant.

William Goodwin, Ph.D., from the University of Glasgow commented on the exceptional discovery. "It is something of a mystery how this child's remains were so perfectly preserved. . . . Normally you only get material with this degree of preservation in material from permafrost areas."[2]

I'm sharing many details here because this landmark discovery was the turning point when it comes to answering the question of where humans fit on the evolutionary family tree.

Using forensic techniques, like the futuristic technology that's depicted in the TV series *CSI*, scientists were able to extract mitochondrial DNA from one of the baby's ribs for analysis. Mitochondrial DNA (mtDNA) is a special form of DNA that's located within the energy centers (mitochondria) inside each of our cells, rather than in the chromosomes, where most of our DNA is found. The reason mtDNA is key when it comes to the question of human evolution is that we inherit it only from our mothers. It's passed from the egg of a mother to both her sons and her daughters, and this typically happens without any of the mutations that can lead to new features in children. This means that the mitochondrial DNA lines in our bodies today are the direct descendants, and exact matches, of the mitochondrial DNA of the woman who began our particular lineage long ago. Because of this unique quality, mtDNA is used to study how people and populations in one place relate to those in other places. It's the uniqueness of this form of DNA that set the stage for the bombshell revealed by the Neanderthal infant.

NOW WE KNOW WHO WE'RE NOT

Using the most advanced techniques, with results that are accepted in the highest courts of law, Russian and Swedish scientists tested the Neanderthal infant's DNA to see how similar hers was to that of modern-day humans. In other words, the scientists wanted to know if the Neanderthal girl was actually one of our ancestors, as the evolutionary family tree leads us to believe. The results of the first studies were published in obscure scientific journals, which concluded, according to the Smithsonian Institution, that "the Neanderthal mtDNA sequences were substantially different from human mtDNA."[3] Although this single statement sounds relatively benign, it's the equivalent of an earthquake with the epicenter right at the root of the human evolutionary tree. Few mainstream news sources shared the discovery, however, and those that did offered the technical details without simplifying them for lay readers or interpreting their significance.

All of that changed, however, in the year 2000. It was then that researchers at the University of Glasgow Human Identification Centre published the results of their own investigation comparing Neanderthal DNA to that of modern humans. The results of their study were shared in a way that made sense even to the most nonscientific reader. And the meaning of what they found could not be dismissed. The conclusion of their report was shared in the peer-reviewed journal *Nature* and directly stated that modern humans "were not, in fact, descended from Neanderthals."[4]

Now there could be no turning back. While scientists had originally believed that the mtDNA of the Neanderthal infant would solve the mystery of our ancestry, it actually did just the opposite.

Key 9: The discovery of an extraordinarily well-preserved female Neanderthal infant—dating back 30,000 years—and the comparison of her mitochondrial DNA to ours, tells us definitively that the earliest modern humans *were not* the descendants of ancient Neanderthals.

NOT YOUR AVERAGE CAVEMAN

If we're not descendants of Neanderthals, then who are our ancestors? Where do we fit on the tree of evolution—do we even belong in Darwin's evolutionary family? The comparison of DNA from Neanderthals and other primate fossils has shed new light on this question. In doing so, however, it's also forced scientists to ponder a new possibility when it comes to unraveling the mystery of our origins.

When I was in school during the 1960s and 1970s, learning about Neanderthals and other prehuman beings such as the *Australopithecus* (the famous Lucy) and the *Homo habilis* (the handy man), we were taught that there was another member of the evolutionary family tree who was a close ancestor as well. In those days, the name used for these distant relatives was the Cro-Magnon. Today, however, that term is no longer used. Paleoanthropologists have replaced it with another that makes more sense, and the reason is self-explanatory. The new name used to identify the beings once known as Cro-Magnons is *anatomically modern humans*, or AMHs.

Scientists generally agree that AMHs first appear in the fossil record approximately 200,000 years ago and mark the beginning of the subspecies *Homo sapiens sapiens*—the term used to describe the people living on earth today.[5] While fossils of bones themselves are more resistant to the elements and can last for millions of years, the DNA found inside bones—in the bone marrow—is much more fragile and typically exists only in relatively recent remains. So although AMHs appeared on earth 200,000 years ago, the oldest DNA discovered from them so far is from a man who lived in Siberia about 45,000 years ago.[6]

In 2003, further advances in genetic technology allowed for the comparison of the earliest anatomically modern human bodies with four newly discovered Neanderthal bodies. A team of European scientists compared the DNA from two AMHs, one that was 23,000 years old and another that was 25,000 years old, with DNA from the remains of the Neanderthals, who were variously dated as living between 29,000 and 42,000 years ago. An article on the findings published in *National Geographic News*, quotes one of the co-authors as saying, "Our results add to the evidence collected previously in different fields, making the hypothesis of a 'Neanderthal heritage' very unlikely."[7] Once again the Neanderthals, often portrayed as primitive cavemen in movies and cartoons, were eliminated as possible ancestors of early modern humans.

Now that we know who our ancestors were not, the focus of paleoanthropology has shifted to discovering who they were. The DNA studies have narrowed the broad field down to one particular candidate. And it's not the candidate supporters of Darwin's theory expected.

THEY ARE US

Scientists now believe that the AMHs are us, and we are they. Any differences between contemporary bodies and those of the AMHs of the past are so slight that they don't justify a separate grouping. In other words, although ancient humans didn't necessarily behave like we do, they *looked* like us, functioned like us, and appear to have had all of the "wiring" in their nervous systems that we have today.

Stated another way, we still look and function as they did 2,000 centuries ago, despite our incredible technological achievements. A 2008 study of AMH remains (still called Cro-Magnon at the time), performed by collaborating geneticists from the universities of Ferrara and Florence in Italy, tells us that these similarities are more than superficial. Researchers report, "A Cro-Magnoid individual who lived in Southern Italy 28,000 years ago was a modern European, genetically as well as anatomically."[8]

It's the fact that members of our species, *Homo sapiens*, haven't changed since our earliest ancestors first appear in the fossil record that poses a problem for the traditional story of evolution, which is based upon slow changes over long periods of time. Discoveries that could not have been made in Darwin's time have shed new light on this lingering mystery.

THE DNA THAT MAKES US DIFFERENT

The set of all human DNA, the *human genome*, was the first DNA sequence of any vertebrate to be entirely mapped. The international effort that made this mapping possible—the Human Genome Project (HGP)—was the result of the largest cooperative biology project in the history of the world.[9] In June 2000, U.K. prime minister Tony Blair and U.S. president Bill Clinton jointly revealed that the first draft of the human code of life had been successfully completed. In doing so, they announced to the world that this unprecedented act of cooperation had opened a new era of lifesaving genetic medicine, and the global industry and economic boom that would follow.

After the success of the HGP, the same techniques used to map human DNA were then applied to other living things. For the first time, scientists could go beyond educated guesses about our genetic relationships and actually compare our life's code to that of any other form of life. The results were nothing short of mind-boggling. While scientists have long known that chimpanzees, for example, are our nearest relatives, for the first time DNA maps allowed them to see just how close that relationship really is.

Genetic mapping revealed that there is only a 1.5 percent difference that separates us from chimpanzees, or conversely stated, we share over 98 percent of the same DNA.[10] When the mapping methods were applied beyond primates, the results were equally astounding. For instance, we share 60 percent of our DNA with a fruit fly, 80 percent with a cow, and 90 percent with a common house cat. We obviously don't look or act like a fly, a cow, or a cat. The big question that comes from such revelations is this: If we have so much in common with other creatures genetically, then why are we so different from them?

The answer to this question goes back to an unexpected discovery made during the HGP: that a single gene can be activated in different ways, and to different degrees, to do different things. What this tells us is that it's not so much about *what* genes we have in common with chimps, cows, flies, and cats. It's more about *how* those genes are activated—or expressed. A gene called *FOXP2,* now understood to be directly linked to our ability to form complex speech, is a perfect example of what I mean here.

FOXP2 is shorthand for Forkhead Box Protein P2, a protein that's involved in the human ability for language. Located on chromosome 7 (precisely at location 7q31), the FOXP2 protein is coded from a gene that has the same name, FOXP2, and is present in both humans and chimpanzees.[11,12] It's obvious, however, that chimpanzees can't sing the Led Zeppelin song "Stairway to Heaven" the way a person can! This fact tells us that there is something more than the gene itself that's involved here. There's something in the way the gene expresses itself that gives us the ability to consistently create the sounds of language. In 2009, a study published in the journal *Nature* gives us a clue as to what that "something" is.

Scientists knew from earlier research that humans and chimpanzees both possess the FOXP2 gene. They'd also determined that the human version of the gene had changed (mutated) at some point in the past, and that the change happened quickly—not slowly and gradually, as the theory of evolution would suggest. Now researchers at the David Geffen School of Medicine at UCLA had determined that this change happened precisely at a critical moment in the unfolding of the human story. According to these scientists, the mutation happened "rapidly around the same time that language emerged in humans."[13] This was a pivotal discovery because for the first time a specific set of mutations in FOXP2 was scientifically linked to our capacity to create complex language.

Additional studies took this research even further and determined when this particular change had happened. According to Wolfgang Enard of the Max Planck Institute for Evolutionary Anthropology, the mutations in FOXP2 that make our complex language possible "happened in the same time frame when modern humans evolved."[14] A *BBC News World Edition* report clarifies this relationship, stating that our capacity for language happened when "changes to two single letters of the DNA code [the representations for the building blocks of amino acids] arose in the last 200,000 years of human evolution."[15]

The speed and precision of the mutations in FOXP2, occurring in just the right two places in the DNA code, are further examples of the kind of change that does not lend itself to the theory of evolution—at least not as we understand the theory today. Why did the changes happen in the way they did? What could have caused just the right shift of DNA letters, at just the right place, within just the right chromosome, to give us the extraordinary ability to share our feelings over a candlelight dinner for two, chant wildly when our team wins the Super Bowl or the World Cup, and whisper into a lover's ear? The best science of the modern world has now given us the answer. The question is, are we willing to accept what the DNA reveals?

FOUND: OUR "MISSING" DNA!

Because humans are classified as the most complex and advanced member of the primate family, it was reasonable for scientists to expect that we would have more chromosomes than our less complex relatives. This is where an unexpected twist in our DNA story begins. Our nearest primate relatives, the chimpanzees, have more chromosomes than we do, with a total of 48 in their overall genome. Ironically, humans have only 46. In other words, it looks like we're *missing* two chromosomes when we're compared to chimps. It's only been recently, using advanced methods of DNA sequencing, that the mystery of "where they went" appears to have been solved. In doing so, however, we once again find ourselves at the threshold of a deeper mystery that holds startling implications!

A closer look at our genetic map shows that our "missing" DNA isn't really missing at all. It's been with us all along; only it's been modified and arranged in a way that wasn't obvious in the past. New research reveals that the second largest chromosome in the human body, forming 8 percent of the total DNA in cells, *human chromosome 2* (HC2), actually contains the smaller "missing" chromosomes found in the chimp genome.[16] In other words, at some point in the past, for reasons that remain controversial, two separate chimp chromosomes got combined into a single larger chromosome that is our chromosome 2.

It's the way these smaller chromosomes combined that may solve the mystery of mutations such as those in FOXP2, and ultimately, the mystery of human origins. While scientists acknowledge that the mutations definitely occurred in FOXP2 and that they happened within the time frame that correlates with the rise of anatomically modern humans, they can't really tell us what caused the change. But they can tell with chromosome 2. And it's this difference that sets chromosome 2 apart.

New technology has revealed precisely what happened to create HC2. I'll share the discovery in two ways with you: first in the scientists' own technical language from the *Proceedings of the National Academy of Sciences* to reveal the discovery itself, and then with a simpler description in lay language to illustrate why this discovery is important to our discussion.

- **The technical explanation.** "We conclude that the locus cloned in cosmids c8.1 and c29B is the relic of an *ancient telomere-telomere fusion* and marks the point at which two ancestral ape chromosomes fused to give rise to human chromosome 2."[17]
- **The simplified explanation.** It appears that long ago two separate chromosomes from chimpanzees (chimp chromosomes 2A and 2B) *merged* or fused into the single, larger human chromosome 2—which is one of the key chromosomes that give us our humanness.

Many of the characteristics that make us uniquely human arise from the DNA fusion that resulted in human chromosome 2. HC2-linked traits include qualities such as our intellect, the growth and development of our brains in general, and specifically the largest part of our brain, the cortex, which is associated with the way we think and act and our capacity for emotion.[18] HC2 contains over 1,400 genes that continue to be mapped and explored today. While a full list, in technical nomenclature, is available through the reference I'm citing in the endnotes, in the following chart I'll share a few simplified examples of these genes to give you a sense of the crucial roles they play in our humanness.[19]

Gene	Influence
Gene TBR1	Key in brain development, particularly the development of the cortex (the largest part of the human brain, which is associated with the way we think and act), our capacity for emotion, empathy, and compassion, and neuron functions (the so-called hardwiring that carries signals within the brain, and throughout the body, to process information)
Gene SATB2	Key in the development of the midbrain and forebrain
Gene BMPR2	Key in osteogenesis (bone tissue formation) as well as cell growth throughout the body
Gene MSH2	Known as a tumor suppressor or "caretaker" gene
Gene SSB	Key in the fetal development of organs that include the heart, brain, eye, kidney, liver, lung, skeleton, spleen, among others

From this small sampling, it's clear that human chromosome 2 plays a significant role in contributing to who, and what, we are. This is especially apparent for the genes TBR1 and SATB2, located on HC2, and the role they play in the development and function of our advanced brain and our extraordinary capacity for emotion. In light of the significance of HC2, the question of how it came into existence becomes more important than ever.

Unlike the previous example of the FOXP2 gene, where changes simply show up in a genome comparison—meaning at one point in time they don't exist in the genetic record of fossils, and at another point in time they do—human chromosome 2 has preserved a record of how it came to exist. It's what this forensic evidence may truly reveal to us that has opened the door to so much speculation. This is where the story of our past takes an unexpected turn, with deeper implications that make our origins begin to sound like the theme of a really good science-fiction novel. You see, the *Proceedings of the National Academy of Sciences* study states that although this kind of fusion is known to happen occasionally, it's rare.

What accompanied the fusion itself opens the door to our new human story.

In the language of the researchers describing this discovery, the fusion was either "accompanied or followed by inactivation or elimination of one of the ancestral centromeres, as well as by events which stabilize the fusion point."[20] While this language is admittedly complex, the message is clear and simple. The study is telling us that during the fusion, or immediately afterward, the overlapping functions from what were originally two separate chromosomes *were either adjusted, turned off, or removed altogether* to make the new single chromosome more efficient.

This fact strongly implies intentionality. And as we've discovered previously, that intentionality led to humanity's possession of many of the extraordinary functions that are found in no other form of life on earth.

Key 10: Human chromosome 2, the second-largest chromosome in the human body, is the result of an ancient DNA fusion that cannot be explained by the theory of evolution as we understand it today.

Two Questions: Why and How?

So now that we know where the missing DNA is located, and how two ancient primate chromosomes were fused into the new larger human chromosome 2, two questions naturally arise:

1. *Why* did this ancient fusion of DNA happen?
2. *How* were the overlapping (redundant) parts of the fusion "switched off" or removed altogether?

The answer to question 1 is that scientists simply don't know. As of this writing, scientists cannot say with absolute certainty why primate DNA got merged in the way that it did, yielding AMHs. While there is certainly no shortage of theories and speculation attempting to explain the mystery, 25 years after this finding was made, the truth is that, at present, there still is no scientific consensus for what could have triggered this miraculous-sounding event.

One thing appears to be certain, however: The DNA that makes us who we are, and what we are, is *not* the result of the process of evolution that Charles Darwin described. My sense is that if we can answer the second question—how the fusion occurred—what we discover will ultimately help us answer the question of why, and much more. When we can answer definitively how the ancient genetic fusion happened and how specific pieces of the fusion were modified so precisely and so quickly 200,000 years ago, the solution to these mysteries will lead us directly to an explanation for why such an extraordinary event took place.

As you may imagine, the discovery of an ancient and complex DNA fusion is interpreted by scientists in different ways. And the differing interpretations have triggered a landslide of controversy. Even after the publication of the article in the *Proceedings of the National Academy of Sciences* described previously, staunch supporters of the evolutionary theory for humans have argued that there are other explanations for the DNA fusion. One theory, for example, proposes that humans and apes, such as chimpanzees and gorillas, all share a common ancestor and that a "split" separated us from them long ago.

If this is true, the chromosome 2 fusion happened to us, and only us, and it happened *after* we had already split from the other primates. They kept their 48 chromosomes and we experienced the fusion that gives us our 46.

This idea makes little sense to me, as it suggests that the DNA that gives us our uniqueness didn't appear until after the uniqueness that caused the split had already happened!

I'm not alone in my thinking, and, to date, evolutionary explanations have not received popular support. I'm sharing an example with you here to illustrate how a radical discovery that attempts to solve one mystery, such as the DNA fusion in chromosome 2, can create even more mysteries as its meaning is digested.

IRREDUCIBLE COMPLEXITY

There is an additional consideration to be made when it comes to the way we think of evolution and the role it may have played in our lives. And while you probably won't see this idea described in classrooms and textbooks (yet), I think it's important to share it here for completeness. The idea is *irreducible complexity.* What this means is much simpler than its name sounds.

I mentioned previously that we have access to knowledge in our era about things that Darwin couldn't have possibly known. It's this fact that makes irreducible complexity worth exploring today. For example, Darwin couldn't have known that even the simplest bacteria, the single-celled *E. coli,* needs 2,000 different proteins to exist; and he couldn't have known that each of those 2,000 proteins has an average of 300 amino acids that make it what it is. The key here is that neither Darwin nor any scientist of the late 1800s or early 1900s could have known just how complex living beings really are. Until recently, nobody could.

Irreducible complexity essentially means that if any portion of a system stops working, the entire system fails. A common mousetrap is often used to illustrate this point. When all of the parts of a mousetrap are in place, it does what it was made to do—what it was *designed* to do: It trips a lever that traps the mouse that has taken the cheese or peanut butter bait, and ends the life of the mouse.

The trap is a system of parts, with each part performing a specific task to accomplish the ultimate goal. For example, there's the lever that holds the bait and there's the powerful spring that comes down with such lethal force when the bait is disturbed that the mouse doesn't even know what hit it. While the trap sounds like a simple gadget, the key is this: *If even one part of the device is missing, the trap simply won't work.* Without the spring, the lever will never snap. Without the lever, the spring will have nothing to trigger. Because all the pieces of the trap are needed for the system to work, it's fair to say that we can't streamline the mousetrap in any way. We can't reduce it to a simpler system and still have the system be functional. It is irreducibly complex.

If we apply this idea to the human body, we see a similar result.

WE ARE LIVING EXAMPLES OF IRREDUCIBLE COMPLEXITY

We all know that when we scrape a knee, the site of the injury will generally bleed briefly and then the bleeding stops. The reason it stops is that blood clots at the site of the scrape. We're so accustomed to seeing this process that it's easy to take the complexity of our blood clotting for granted. We just assume that it's going to happen. And the fact that it does is a perfect example of irreducible complexity. When we scratch, cut, or break our skin, 20 separate proteins must be already in place and ready to act for our blood to clot and the bleeding to stop.

This fact is key to our discussion of irreducible complexity for one important reason: *If even one of the 20 proteins needed for clotting is missing, the bleeding will continue.* Whether we wait 10 minutes or 10 hours, the result will be the same. Our blood can only coagulate when all the proteins that make clotting possible are in place.

Our blood's ability to clot is an example of a life function that could not have developed through evolution. To do so, 20 proteins would have already needed to be formed and in the same place before the blood that gives our bodies life could form. If these components had not already been in place, our ancestors would have bled to death with the first minor injuries they sustained—meaning we might not be here, because they might have died without producing offspring. And this is only one example.

Here's another. The little waving arms (cilia) that allow cells, including sperm cells, to travel in fluid have more than 40 moving parts that must all be present for the cilia to wave. If any part is missing, the cells can't move. If ancient sperm cells from a male of our species had not been immediately empowered to "swim" toward the egg of a female, reproduction could not have occurred.

And there's more.

The human cell has been called the single most complex piece of machinery ever known to exist. Until the mid-20th century or so, cells were essentially thought of as tiny bags of salt water holding dissolved elements. We now know that nothing could be further from the truth. In fact, if we could enlarge a single cell to the size of a city, we would discover that the cell is more complex than just the infrastructure that keeps it going. A sample of a cell's important structures includes:

- Ribosomes that manufacture proteins.
- Endoplasmic reticulum that makes and transports important chemicals used by the cell.
- A nucleus that carries instructions for the cell on how to function.
- Microtubules that allow the cell to move and change shape.
- Cilia (little waving arms) that allow some cells to move in fluid.
- Mitochondria that generate energy for the cell.

- A membrane that communicates with the environment and determines what gets into and leaves the cell.

This is just a sampling of the myriad processes that are happening at any given moment in each of the approximately 50 trillion cells of the human body. As we discover what each process does, it becomes obvious that all this cellular machinery had to be already created, and in place, for our earliest cells to do what they do. From clotting blood to swimming cilia, the body holds many examples of irreducible complexity.

To even the most skeptical scientist, it's obvious that the DNA of life is based upon structure, order, and the sharing of information that tells our cells what to do and when to do it. In nature, this kind of order is often seen as a sign of intelligence.

Key 11: The 20 proteins that make the clotting of blood possible and the 40-plus components of the cilia (wiggly tails) that allow cells to move through a fluid are just two examples of functions that could not develop gradually over a long period of time as evolution suggests. In both examples, if even one protein or component part is missing, the function of the cells is lost.

In candid interviews late in his life, Albert Einstein shared his belief that an underlying order of information exists in the universe, as well as his sense of where that order comes from. During one of those conversations, he confided, "I see a pattern but my imagination cannot picture the maker of the pattern. We all dance to a mysterious tune, intoned in the distance by an invisible piper."[21] In our search for human origins, the very presence of the order and intentionality we see in our DNA is a sign that Einstein's invisible piper exists.

WE'RE OVER-ENDOWED!

There is an additional theme of evolution theory that I have intentionally waited until now to mention. It's a corollary to Darwin's theory, first stated by a colleague and fellow supporter of Darwin, British naturalist Alfred Russel Wallace. Through his work, Wallace defined the evolutionary principle that paves the way for the rest of this book. Building upon Darwin's original work, Wallace made an extraordinary observation when it comes to the development of new features in a species. I'll share Wallace's corollary, stated in his own words, and then apply his statement to what we now know about our own development.

In the final chapter of his book *Contributions to the Theory of Natural Selection,* published in 1870, Wallace leaves no doubt with his readers about what he's saying: "Natural Selection would only have endowed savage man with a brain a little superior to that of an ape, whereas he actually possesses one very little inferior to that of a philosopher."[22] In this somewhat complex passage, Wallace is stating that nature only gives us what we need, when we need it, and does so through evolution, which Darwin defined as a slow and gradual process. In other words, the theory says that we have abilities such as standing upright, advanced peripheral vision, and the ability to share our emotions through smiles, frowns, and other facial expressions because we needed them at some point in the past.

Herein lies the problem. We're all over-endowed! And it appears that we have been since the dawn of our existence.

Key 12: Humans appeared on earth with the same advanced brains and nervous systems we have today and with the ability to self-regulate vital functions already developed, contradicting the corollary to evolution theory that nature doesn't "over-endow" with such features until they are needed.

THE NEW HUMAN STORY

Following 150 years of the best human minds applying themselves under the auspices of the world's most respected universities, being funded with tremendous sums of money, and using the most sophisticated technology available to solve the mystery of our origins, if we were on the right track, it would seem that we'd be farther along than we are today. In light of the failure of Darwin's theory to explain our existence, and in consideration of the new evidence that I've presented, it's reasonable to ask the question that's become the big pink elephant in the room: What if modern science is on the wrong track?

What if we're trying to prove the wrong theory and writing the wrong human story? The answer to this question is the reason I've written this book. If we're on the wrong track, it may help to explain why so many of the solutions applied to the world's problems aren't working. This would mean that our thinking and the "solutions" our approaches have produced are based on something that's not true!

Why not allow the evidence to *lead* us to the story of our past, rather than trying to *force* the evidence into a template that was formed over a century and a half ago? If we're serious about solving the deepest mystery of our existence, it would make sense for us to open our minds and allow for another interpretation of the data we've collected during a century and a half of study.

What if there is no evolutionary path leading to modern humans? What if the pieces of the genetic puzzle that makes us who we are were suddenly locked into place all at once rather than accumulating gradually over time? What would such a story look like? The data from studies of human chromosome 2 and other DNA studies, the lack of fossil evidence documenting the transition from one hominid species to another, and the lack of common DNA between humans and less advanced primates all suggest that we may not belong on the same tree with the early hominids commonly shown in the textbooks. In fact, they suggest that we may not belong to a tree at all! The evidence suggests that our history may be represented best as a stand-alone shrub—an evolutionary bush—that begins and ends with us.

In other words, we may find that we're a species unique unto ourselves.

Key 13: A growing body of physical and DNA evidence suggests that our species may have appeared 200,000 years ago with no evolutionary path leading to our appearance.

This is not to say that evolution doesn't exist or hasn't occurred anywhere. It does and it has. As a geologist, I've seen firsthand the fossil record of the evolution that's occurred in a number of other species. It's just that when we attempt to apply what we know of the evolution of plants and animals to humans, the facts don't support the theory. They fail to explain what the evidence reveals.

If we were to place the essence of the new discoveries about us into a concise list, the statements that follow would offer a high-level summary. Additionally they would give us a good idea of where the new theories, and our new story, may be heading.

HERE'S WHAT WE'RE NOT

- The theory of living cells evolving (mutating randomly) over long periods of time *does not,* and *cannot,* explain our origins or the complexities of our bodies.
- The evolutionary family tree for humans *is not* supported with physical evidence.
- DNA studies prove that we *did not* descend from Neanderthals, as previously believed.
- We *have not* changed since the first of our kind, the anatomically modern humans, appeared in the fossil record of the earth approximately 200,000 years ago.
- The precise events that produced the DNA that gives us our uniqueness *are not* commonplace in nature.

So now that we know what we're *not,* what does the best science of our time tell us about who we *are*? What does the new human story look like?

HERE'S WHAT WE ARE

- AMHs appeared on earth approximately 200,000 years ago with the DNA and the advanced brain and complex nervous system that set us apart from other forms of life already formed and functioning.
- We appear to be a species unique unto ourselves, with our own simple family tree, rather than being a variation of preexisting forms of life traditionally shown on an increasingly crowded family tree.

- The DNA that makes us unique is the result of a rare arrangement of chromosomes, which are fused and optimized in a way that cannot be identified as random.

Key 14: An honest scientist, who is not bound by the constraints of academia, politics, or religion can no longer discount the new evidence about our human origins and still remain credible.

In the course of my life, I've discovered that when I find something that makes no sense to me, it's generally because I don't have all the information. I believe that the conventional scientific theory of human origins—the story we've been asked to accept—falls into this category. The evidence that I've shared in this chapter clearly doesn't support Darwin's story of evolution. While the science is good and the methods scientists use are sound ones, it's our responsibility to recognize the limits of what science can reveal. As I mentioned previously, while scientific evidence can definitely tell us *what* has happened in the past, it cannot necessarily tell us *why* something has happened or if conscious intention led to the event.

For example, when we see a fire brightly burning on a warm summer night in the middle of a grassy field, scientific knowledge tells us that a spark of some kind has ignited that fire. It's telling us that a fire can only come from a) a source of heat great enough to start the fire (the *kindling temperature*), or b) another fire, such as the accidental spark of a lawnmower blade hitting a rock, the intentional spark of a match or a cigarette lighter, or the natural spark of lightning striking the ground. My point here is that without first knowing the circumstances that were in place when the fire began, science can't tell us the precise reason that the spark occurred, or if it was an intentional act. If a fire occurred hundreds or thousands of years in the past, much of the evidence related to its circumstances would be lost in the fog of time. All we would know from the charred remains of a log or a scorched rock is that there had been a fire.

The fusion of DNA in human chromosome 2 is like that fire in the field. Science can tell us that the fusion making it possible occurred and how that fusion occurred. But because scientists can't determine all of the circumstances surrounding the fusion—as these have been lost over the ages—we're left to rely upon facts, logic, and deductive reasoning to make sense of what we see. The same point that I'm making here for our chromosome 2 can be made for our gene FOXP2.

WE'RE HUMAN BY DESIGN

I want to be absolutely clear that what I say next is not the conclusion of peer-reviewed science, although I've spoken with mainstream scientists who have told me that they suspect it is true, yet are reluctant to speak publicly about their suspicions for fear of losing their reputation, their credibility, and even their jobs. When I honestly consider the evidence that I've shared in these chapters, it simply makes sense to look beyond evolution and an unbelievably good run of biological "luck" to explain the fact of our existence.

The evidence overwhelmingly suggests that:

1. **We are the result of an intentional act of creation.**
    - The mutations in FOXP2 and human chromosome 2 are precise
    - The mutations in FOXP2 and human chromosome 2 appear to have happened quickly rather than through a long, slow evolutionary process

- The optimization of human chromosome 2 that occurred *after the fusion* appears to be intentional
- After 150 years of searching, the fact that no physical evidence has been discovered to link us to other forms of life on the tree of primate evolution suggests that we may be a species unto ourselves, with no evolutionary history

2. **We are the products of an intelligent form of life.**
    - The timing, precision, and accuracy of our genetic mutations, and the technology required to yield such mutations, implies the forethought and intention of an advanced intelligence
    - The intelligence that carried out the genetic modifications giving us our humanness had the advanced technology to do 200,000 years ago what we are only learning to do today (for example, DNA fusion and gene splicing)

To honestly acknowledge these possibilities opens us to a paradigm that shifts the way we feel about ourselves and view our place in the universe. With this shift, we free ourselves from a paradigm of lonely insignificance and move into one of possessing a rare heritage that we are only beginning to explore. And that's where this book begins. We're here with the bodies and the nervous systems that afford us the abilities of compassion, empathy, intuition, self-healing, and much more. The fact of their presence within us suggests that we're intended to utilize—and master—the sensitivities that we arrived with.

The new human story begins with our beginnings. It begins with the fact that from the time of our origin we've been neurologically wired for extraordinary abilities. This design affords us extraordinary ways of living and extraordinary lives.

The question that immediately comes to mind when we consider that we've had such advanced characteristics from our beginning is this: How do we fully awaken these capabilities in our lives today? In the chapters that follow, I invite you to share a journey of discovery in which we do our best to answer this question and explore what it means to be human by design.

chapter three

The Brain in the Heart

Heart Cells That Think, Feel, and Remember

"If the 20th century has been, so to speak, the Century of the Brain, then the 21st century should be the Century of the Heart."

— Gary E. R. Schwartz, Ph.D., and Linda G. S. Russek, Ph.D.

The first fossils of anatomically modern humans were discovered under a rock ledge in southwestern France in 1868. The name given the formation where the discovery was made is *abri de Cro-Magnon* (meaning, in the local dialect, "shelter of the cave-dwelling Magnon family"), which was soon shortened to Cro-Magnon.[1] This location became the namesake for Cro-Magnon humans, now known as AMHs. Regardless of the name we use to describe the early people who lived in this region of France, these ancient humans were different from any other form of life that existed at the time or has existed since.

Just as forensic scientists today are able to use computers to reconstruct the muscle mass, flesh, and facial features of a modern human body that's been reduced to a skeleton, scientists have been able to use the same technology on AMH skeletons as well, and the features they've been able to reconstruct look like ours—because they are us! The archaeological and DNA evidence tells us that we haven't changed for 200,000 years.

Anatomically modern humans had features that set them apart from other ancient beings, such as the Neanderthals, whom we now know lived at the same time. AMH males, averaging approximately five feet nine inches,[2] were tall in comparison to Neanderthal males, whose height ranged from five feet four to five feet five.[3] AMH bone structure was thinner and more delicate overall, their skulls were more rounded in the back, and their faces were smaller, with more pointed chins.

In addition to these visible differences, AMHs had advanced biology—differences that couldn't be seen with the naked eye that gave them an edge over all other forms of life on earth. Many scientists attribute their survival through the last ice age into modern times to these advanced features, which include a brain 50 percent larger than that of their nearest primate relative; a complex language; an anatomy that enables them to stand, walk, and run upright; and opposing thumbs and fingers.

For clarity, I want to reemphasize that the makeup of the AMHs of 200,000 years ago has been determined to be essentially the same as that of humans today, both in genetics as well as physiology. Because of this, the assumption is that the advanced features we have today were also part of our human ancestors. Their inherent features would have included the ability we have today to tap the network of neurons, vital organs, and glands throughout the body to trigger their extraordinary potentials in a conscious way—and do so at will—to experience benefits such as deep intuition and self-healing.

I'm contrasting the presence of this network in AMHs to other forms of life that have neural networks as well, yet are less developed and must rely upon something in their external environment to trigger the benefits of their biology. A small zebra fish, commonly used in laboratory experiments, is a perfect example of what I mean here. It's only when the fish is stimulated by something outside of its body, such as a visual cue that makes the fish think it's drifting backward in a current, that 80 percent of the neurons in its brain fire all at once. This is the equivalent of signaling *All systems go!* in the fish's body. It's this simultaneous triggering of neurons that gives the fish immediate access to the benefits of such a coherent experience. In this instance, the zebra fish is able to tap the combined neural power to swim quickly and correct its course.[4]

Ancient humans had the ability to access their neural power without the need for an external signal as well. They could trigger their potent network of specialized cells and organs on demand. And we continue to have that ability today.

This is where the new human story that our biology is revealing to us departs from Darwin's original ideas about evolution. Having conscious access to our advanced neural network gives us the godlike powers of intuition, self-healing, super-consciousness, and much more. These benefits have been utilized by yogis and shamans throughout the ages and described in their sacred mystical texts. Perhaps it's not surprising that the key to accessing such advanced features of our experience begins with our mastery of the single organ that has been the focus of our ancestors' teachings for millennia: the human heart.

A recent discovery within the heart is shaking the foundation of what we've been led to believe about the heart's role when it comes to us and our bodies. Interestingly, while the discovery is overturning traditional thinking when it comes to which organ we consider to be the master organ of the body, it actually parallels teachings found in our most ancient and cherished traditions.

THE UNCHARTED HEART

When average people are asked to identify the organ that controls the key functions of the body, more often than not the answer is the same. They'll say it's the brain. And it's no surprise that they do. From Leonard da Vinci's day, 500 years ago, until as recently as the late 1990s, people throughout the Western-educated world have believed that the brain is the conductor leading a symphony of functions in the body that keep us alive and well.

It's what we've been taught. It's what we've been led to believe. It's what teachers have stated with authority. It's the premise that doctors and health-care workers have based life-or-death decisions upon. And it's what most people will say when asked to identify the roles of the most important organs of the body. The belief that the brain is the master organ of the human body has been embraced and endorsed by the some of the most innovative scientists and thinkers at the most highly esteemed institutions and universities in modern history, and it persists in mainstream thinking today.

The home page on the website for the Mayfield Clinic, affiliated with the University of Cincinnati's neurosurgery department, is a beautiful example of this way of thinking when it comes to the brain. It reads:

> The brain is an amazing three-pound organ that controls all functions of the body, interprets information from the outside world, and embodies the essence of the mind and soul. Intelligence, creativity, emotion, and memory are a few of the many things governed by the brain.[5]

The belief that the brain is the control center for the human body, our emotions, and our memories has been so universally accepted that it's been taken for granted almost without question for a long time—that is, until now. As the discoveries described in the following chapters will reveal, this perspective is only one piece of a much bigger story.

Today, what we thought we knew about the brain is changing. It has to. The reason is simple: The discoveries described in this chapter, and the decades of research that have followed, tell us that the brain is only part of the story. While it's certainly true that the brain's functions include things like perception, motor skills, information processing, providing chemical triggers for every urge we feel automatically—including fatigue, hunger, and sexual desire—and also maintaining the strength of our immune system, it's also true that the brain can't do these things alone. The brain is only one part of the bigger picture that's still emerging and largely untold. It's a story that begins in the heart.

Key 15: As part of our advanced nervous system, the heart partners with the brain as a master organ to inform the brain of what the body needs in any given moment.

THE HUMAN HEART: MORE THAN JUST A PUMP

When I was in school, I was taught that the main purpose of the heart is to move blood through the body. I was told that the heart is a pump—an amazing pump, yet still a pump, plain and simple. I was also taught that the heart has one job, which is to keep blood moving over the course of our lifetime. By any measure this is an extraordinary accomplishment, as the adult heart beats an average of 101,000 times a day. As it does so, it circulates approximately 2,000 gallons of blood through 60,000 miles of arteries, capillaries, veins, and other blood vessels![6]

A growing body of scientific evidence now suggests, however, that the pumping of the heart, as important as this function is, may pale in comparison to the additional functions of the heart that have only recently been discovered. In other words, while the heart *does* pump blood powerfully and efficiently through the body, pumping may not be its primary, or exclusive, purpose.

For thousands of years, our ancestors regarded the human heart as the center of thought, emotion, memory, and personality—the true master organ of the body. Traditions to honor the role of the heart were created and passed down through generation after generation. Ceremonies were performed and techniques developed to utilize the heart as a conduit of intuition and healing.

The heart is mentioned 830 times in the Bible, and the word *heart* appears in 59 of the Bible's 66 books.[7] The book of Proverbs describes the heart as a source of vast wisdom that requires a cultivated understanding to make sense of it: "Counsel in the heart of a man is like deep water; but a man of understanding will draw it out."[8]

The same sentiment is stated clearly in the native wisdom of the North American Omaha people, whose tradition invites us: "Ask questions from your heart and you will be answered from the heart."[9]

The Lotus Sutra of the Mahayana Buddhist tradition teaches of the "hidden treasure of the heart."[10] This treasure is described in the scripture as being "as vast as the universe itself, which dispels any feelings of powerlessness."[11]

Key 16: Ancient traditions have always held that the heart, rather than the brain, is the center of deep wisdom, emotion, and memory, as well as serving as a portal to other realms of existence.

Clearly references such as these are referring to the heart as something much more than a physical pump. They're telling us, just as the visionary philosopher Rudolf Steiner, the creator of the Waldorf method of education, and Harvard University's biodynamic agriculture scientist John Bremer suggested to Harvard Medical School students early in the 20th century, that there's something more to the heart than we've been led to believe.[12]

If we're willing to embrace what the following discoveries tell us, we can agree with Steiner and Bremer that our hearts are capable of something much more mysterious, powerful, and beautiful than simply being a pump.

Our exploration to know ourselves has created a journey that swings like the extremes of a pendulum. From the time my life began in the early 1950s until today, I've seen the pendulum of thinking swing from an extreme view of the heart as an isolated pump that can be serviced and replaced like a machine back toward a balanced view that the heart is much more than just a pump. There is a new recognition of the heart as an integral source of memories, intuition, and deep wisdom, as well as a biological organ that gives us life. This shift of views invites us to rethink which organ we can honestly call the master organ of the body.

THE "LITTLE BRAIN" IN THE HEART

In 1991, a scientific discovery published in the journal *Neurocardiology* put to rest any lingering doubt that the human heart is more than a pump. The name of the journal gives us a clue to the discovery of a powerful relationship between the heart and the brain that went unrecognized in the past. A team of scientists led by J. Andrew Armour, M.D., Ph.D., of the University of Montreal, which was studying this intimate relationship between heart and brain, found that about 40,000 specialized neurons, or *sensory neurites*, form a communication network within the heart.[13]

For clarity, let me say that the term *neuron* describes a specialized cell that can be excited (electrically stimulated) in a way that allows it to share information with other cells in the body. While large numbers of neurons are obviously concentrated in the brain and along the spinal cord, the discovery of these cells in the heart and other organs, in smaller numbers, gives new insight into the profound level of communication that exists within the body.

Neurites are tiny projections that come from the main body of a neuron to perform different functions in the body. Some carry information *away* from the neuron to connect with other cells, while others detect signals from various sources and carry them *toward* the neuron. What makes this discovery exceptional is that the neurites in the heart perform many of the same functions that are found in the brain.[14]

In simple terms, Armour and his team discovered what has come to be known as the *little brain* in the heart, and the specialized neurites that make the existence of this little brain possible. As the scientists who made the discovery say in their report, "The 'heart brain' is an intricate network of nerves, neurotransmitters, proteins, and support cells similar to those found in the brain proper."[15]

Key 17: The discovery of 40,000 sensory neurites in the human heart opens the door to vast new possibilities that parallel those that have been accurately described in the scriptures of some of our most ancient and cherished spiritual traditions.

A key role of the brain in the heart is to detect changes of hormones and other chemicals within the body and communicate those changes to the brain so it can meet our needs accordingly.

The heart's brain does this by converting the language of the body—emotions—into the electrical language of the nervous system so that its messages make sense to the brain. The heart's coded messages inform the brain when we need more adrenaline, for example, in a stressful situation, or when it's safe to create less adrenaline and focus on building a stronger immune system.

Now that the little brain in the heart has been recognized by researchers, the role it plays in a number of physical and metaphysical functions has also come to light. These functions include:

- Direct heart communication with sensory neurites in other organs in the body
- The heart-based wisdom known as *heart intelligence*
- Intentional states of deep intuition

- Intentional precognitive abilities
- The mechanism of intentional self-healing
- The awakening of super-learning abilities
- And much more

The heart's little brain has been found to function in two distinct yet related ways. It can act:

- Independently of the cranial brain to think, learn, remember, and even sense our inner and outer worlds on its own[16]
- In harmony with the cranial brain to give us the benefit of a single, potent neural network shared by the two separate organs[17]

Armour's discovery has the potential to forever change the way we think of ourselves. It gives new meaning to what's possible in our bodies and what we're capable of achieving in our lives. In his words: "It has become clear in recent years that a sophisticated two-way communication occurs between the heart and the brain, with each influencing the other's function."[18]

The science from the new field of neurocardiology is just beginning to catch up with traditional beliefs when it comes to explaining experiences such as intuition, precognition, and self-healing. This is especially apparent when we examine the principles offered in some of our most ancient and cherished spiritual traditions. Almost universally, historical teachings demonstrate an understanding of the heart's role at the level of having direct influence upon our personalities, our daily decisions, and our ability to make moral choices that include the discernment of right and wrong.

The Coptic Christian saint Macarius, founder of an ancient Egyptian monastery that bears his name, powerfully captured these levels of potential within the heart in saying:

> The heart itself is but a little vessel, and yet there are dragons, and there are lions, and there are venomous beasts, and all the treasures of wickedness; and there are rough uneven ways, there chasms; there likewise is God, there the angels, there life and the kingdom, there light and the apostles, there the heavenly cities, there the treasures, there are all things.[19]

Among the "all things" that St. Macarius described, we must now include the new discoveries that document the ability of our hearts to remember life events—even when the heart is no longer in the body of the person that experienced the events.

MEMORIES THAT LIVE IN THE HEART

One of the mysteries of heart transplants is that an undamaged heart will continue to beat after being removed from its original owner—sometimes for a period of hours—and is able to resume functioning after it's placed in a new body and connected to new blood vessels and nerves. The crux of the mystery is this: If the brain were truly the master organ of the body, responsible for sending instructions *to* the heart telling it to beat and pump blood, then wouldn't the heart stop beating and functioning after it had lost its connection to the brain? Why does it function without those instructions?

The following true-life accounts, and the discovery they led to, shed a powerful light on the mystery of the heart and offer new insight into its deeper role in our everyday lives.

The first successful human heart transplant took place in Cape Town, South Africa, on December 3, 1967. On that day, Christiaan Barnard, M.D., placed the heart of a 25-year-old woman who had been in a fatal automobile accident into the body of Louis Washkansky, a 53-year-old man with a damaged heart.[20] From a medical standpoint, the procedure was an overwhelming success. The woman's heart immediately began functioning within the man's body, just as the transplant team had expected.

One of the biggest hurdles in all transplants, including Washkansky's, is that the immune system of the person receiving the heart (or any organ for that matter) does not recognize the new organ as its own and tries to reject the foreign tissue. For this reason, doctors use specialized drugs to suppress the recipient's immune system and trick the body into accepting the new organ. The good news is that this technique is successful in reducing the chances of rejection. The success comes with a high price, however.

With a severely weakened immune system, the recipient of a new organ becomes susceptible to infections such as common colds, flu, and pneumonia. And this is precisely what happened with the world's first human heart transplant. Although Louis Washkansky's new heart functioned perfectly until his last breath, he died 18 days after the transplant from complications of pneumonia. His survival with a new heart for over two weeks demonstrated, however, that an organ transplant was a viable possibility in cases where an otherwise healthy body loses an organ to accident or disease.

In the decades that have followed Barnard's first transplant, the procedures have been perfected to the point that human heart transplants now happen on a routine basis. In 2014, approximately 5,000 heart transplants were performed worldwide.[21] And while that number sounds high, when it is compared to the list of 50,000 people waiting for a new heart from a compatible donor, it's clear that the demand for organ donors will remain high for the foreseeable future.[22]

The reason that I'm sharing the background of heart transplants here is because it has a direct connection to the theme of this chapter. From the time of the first procedures, there has always been a curious phenomenon that's now acknowledged in the medical community as a possible side effect of a heart transplant. It's called *memory transference.* One of the earliest examples of this phenomenon was documented by the direct experience of a woman named Claire Sylvia who received a transplant in 1988. Her memoir, *A Change of Heart*, is an account of her experiences as a recipient and how they opened the door to researchers for the serious study, and eventual acceptance, of how life's memories can be preserved within the heart itself, regardless of whose body the heart resides within.[23]

Sylvia, once a professional dancer, had successfully received the heart, as well as the lungs, of a donor whose identity was not initially disclosed. Not long after her operation, she began to crave foods that she'd never been especially drawn to in the past, such as chicken nuggets and green peppers. And when it came to the nuggets, the cravings were very specific. Sylvia found herself drawn unexplainably to the restaurant chain KFC to satisfy her cravings. As she had never developed a taste for these kinds of foods before her surgery, her friends, family, and doctor were mystified by the cravings.

Just before her operation, she'd been told that she was receiving the organs of a young man who had died in a motorcycle accident. Although the details of donors are generally not shared with the person receiving their organs, Sylvia followed up on the information she had and discovered the identity of the young man in a local obituary, along with the address of his parents. It was during a visit she had with them that Sylvia learned some of the details about the life of their son, Tim, whose heart and lungs were now in her body. And those details confirmed for her mind what she already sensed intuitively to be true: Tim loved precisely the kind of chicken nuggets and green peppers that she was now craving. It was clear that Tim's desire for the foods he enjoyed in his lifetime was now part of Sylvia's experience and that she had acquired her cravings through memory transference.[24]

Key 18: Scientific documentation of memories carried from a donor into the body of a recipient through the heart itself—*memory transference*—demonstrates just how real the heart's memory is.

While Claire Sylvia's story is one of the earliest and best-documented accounts of memory transference through a heart transplant, there have been additional examples since. In each one, there is a change in the personality of the person who gets a new heart. These changes range from a new preference for specific foods to differences in personality and even sexual orientation that reflect the preferences and personality of the donor.

And while examples of personality changes are fascinating, the stories don't end there. The emotional memories of our lives appear to be so deeply ingrained in the heart's memory that they are preserved with tremendous clarity and are commonly reexperienced by the person receiving the heart in a transplant.

While skeptics of heart memory theories have proposed a number of alternative explanations for the post-transplant changes in personality and lifestyle, including drug reactions and subconscious influences, there's a particular kind of experience that cannot be explained away by the skeptics' theories. It's this kind of documented case history that has led to the acceptance of memory transference as a fact of life rather than a curious coincidence.

IF THE HEART IS ALIVE, THE MEMORIES REMAIN

Two years after Claire Sylvia's book came out, in 1999, Paul Pearsall, M.D., a neuropsychologist, published another pioneering book documenting case histories of the heart's memory. This book, *The Heart's Code,* included true-life accounts of the memories and dreams, even nightmares, experienced by people who had undergone heart transplants. What made one of the accounts so extraordinary is that the recipient's experiences could be confirmed as factual events that had happened in the donor's lifetime. This case involved an eight-year-old girl who received a heart from another girl two years her elder.

Almost immediately after the surgery, the young girl began having vivid and frightening dreams—nightmares—of being chased, attacked, and killed. While her transplant was a success technically, the psychological impact of the nightmares continued. She was eventually referred to a psychiatrist for a mental health evaluation. The events and images that the young girl described were so clear, consistent, and detailed that the psychiatrist was convinced the dreams were more than curious by-products of the transplant. She felt sure that the girl was describing memories of a real-life experience. The question was, from whose memory?

The authorities were eventually brought in on the case and quickly discovered that the girl was recounting the details of an unsolved murder that had occurred in their town. She could share the specifics of where, when, and how the murder happened. She could even repeat the words that were spoken during the attack and name the murderer. Based upon the details she provided, police were able to locate and arrest a man that fit the circumstances and description. Eventually he was put on trial and convicted of the assault and murder of the ten-year-old girl whose heart was now in the body of the eight-year-old.[25]

This account tells us how real the little brain in our hearts is, and how it can function in ways that were once believed to happen only in the cranial brain. The discovery of this second brain in the heart, and the compelling evidence of its ability to think and remember, has opened the door to a vast array of possibilities in our lives.

What does the hidden potential of the heart mean for our lives? Since the time Leonardo da Vinci first diagrammed the nerves connecting the brain to the major organs in the body nearly 600 years ago, we've been led to view the heart and brain from an either/or perspective.[26] Scientists, engineers, and analytical problem solvers have long felt that the brain is the master control center for the functions in the rest of the body, so they've often discounted the heart. At the same time, artists, musicians, and intuitive thinkers have typically felt that the heart is the key to inspiration, insight into life challenges, and the deep wisdom that can guide our lives, and have readily discounted the thinking capacity of the brain in these roles. It's now apparent why this either/or thinking generally doesn't work out so well.

To separate the brain from the heart gives us an incomplete picture of our full potential. Clearly, the more we discover about the way the heart and brain can function as a single network to regulate the body, it becomes clear that our greatest benefit comes from harmonizing both organs to work together rather than focusing on one or the other exclusively. The more we understand about how to create heart-brain harmony, the more we can use this understanding to tap the power of our greatest potentials!

Twentieth-century playwright and congresswoman Clare Boothe Luce once said, "The height of sophistication is simplicity."[27] The truth of her words applies especially to nature. Nature is simple and elegant until we make it difficult through awkward descriptions and complex formulas. So what could be simpler than the brain in our hearts and the brain in our heads naturally forming a single, potent network that enables us to experience deep intuition, empathy, and compassion in our lives?

While these sorts of remarkable states of consciousness are typically attributed to the extraordinary skill of trained mystics, monks, and yogis, my sense is that these are actually ordinary states of consciousness available to each of us that our culture has simply forgotten.

HEART WISDOM IN EVERYDAY LIFE

Have you ever been faced with a decision that seemed impossible to make? Maybe it was a question of whether to go forward with a medical procedure that wasn't aligned with your belief system.

Maybe it was whether to stay in a difficult relationship or end it. Maybe if you answered the question the wrong way, you or a loved one would face life-or-death consequences.

As different as these issues are from one another, the thing that links the decisions is that none has an absolute answer. For each situation there is neither right nor wrong. There is no "book of truth" you can turn to when facing tough decisions that will tell you which option is best. And if you've ever been in a situation where you had to make this kind of decision, you probably discovered that every friend you ask for help has a unique opinion as to why a certain path is right for you, so you end up with a collection of opinions that make the original choice even more confusing.

Or maybe something else happened. Maybe you did follow the recommendation of a close friend or relative who meant well in trying to help. Maybe you tried the age-old solution of answering your question by making a list of pros and cons. This was my mother's advice for how to handle every tough decision when I was growing up.

"On a sheet of paper, make two columns," she would say. "Title one column 'Pros,' for the good things about your choice, and the other column 'Cons,' for the things that are not so good. Then add up the pros and cons and you'll have your answer. And if that doesn't work, go ask your father."

I can tell you from experience that neither of these solutions works. Before he left our family when I was 10 years old, my father was largely unavailable when it came to the big questions of life. So if my mother couldn't answer my question, I had few options. And the list that my mother would ask me to make always seemed to be skewed in favor of what I *wanted* the answer to be, rather than what the best answer really was.

The reason it's so difficult to make big decisions that have no clear answer is directly linked to the way we've been conditioned to think. Most of us have been trained to think exclusively with our brains. And while there are times when it definitely serves us to use mental reasoning, such as when we're creating step-by-step plans to build a house, solve a complex mathematical problem, or lay out steps leading to income security in the future, there are times when we actually limit ourselves by trying to answer life's big questions through reasoning alone. Solving our problems through reasoning alone can sometimes be a slow and cumbersome process for two main reasons:

- **Thought-based choices are typically filtered through our perceptions and past experience.** When it comes to choosing our role in an intimate relationship, for example, our logic-based decision is made through the filters of our self-image. This is why our answer to the question *Who am I?* is so important. Our minds will make the choice to continue a relationship, or to end it, through the lens of our sense of personal worth. As we'll see in the next chapter, that sense is, in part, derived from the scientific story of evolution and the feeling of insignificance it gives us.
- **Our minds tend to justify the answers we arrive at using circular reasoning, a way of thinking that supports a conclusion by restating it.** For example, if I were to say to you, "I like Bon Jovi because it's my favorite band," the circular part of the reasoning is shown by my stating the same thought twice, using the words *like* and *favorite*. With them, I would be using the second thought to justify the first, and the first thought to justify the second.

  This kind of reasoning can play out in unexpected ways, such as reinforcing our fear of taking a chance with a new and challenging job that's been offered to us and justifying turning it down. In such an instance, the circular logic goes something like this: *I already have a safe position in a good company* ➔ *If I accept the new job and the new responsibilities, I may not be able to meet the expectations that come with it* ➔

If I lose the new job I'm not safe ➔ *I already have a safe position in a good company.*

To be clear, I'm not suggesting that any of the preceding characteristics of mental problem solving is bad, or even good, for that matter. What I am saying here is that there are different kinds of challenges in life that are best solved through different ways of thinking—some with the brain and some with the heart. And while heart-based thinking may be less familiar in our fast-paced world of technology and digital information, in a very real sense heart-based wisdom is perhaps the most sophisticated technology we could ever have at our disposal.

Rather than thinking through the pros and cons of a decision, or weighing the odds that an experience from the past will repeat itself in the present, our heart intelligence knows instantly what's true for us in the moment. Regardless of whether we choose to accept or ignore the heart's wisdom, it's there for us. This is true when it comes to how we feel about other people, as well as when we're making important choices in our lives. Scientific studies into the accuracy of our first impressions when it comes to trusting another person are a perfect example of the heart's wisdom that we've all experienced at some point in our lives.

THE HEART KNOWS IMMEDIATELY

A study led by Alex Todorov, Ph.D., a Princeton University psychologist, showed that when we meet another person for the first time, our assessment of that person is almost immediate. "We decide very quickly whether a person possesses many of the traits we feel are important, such as likeability and competence, even though we have not exchanged a single word with them," he says. "It appears that we are hard-wired to draw these inferences in a fast, unreflective way."[28]

When we think about how quickly we form our opinions of people we've never met, it actually makes perfect sense. It's nature's way of keeping us safe. Our ancestors, for example, didn't have the luxury of visiting for hours to get to know the people they came face-to-face with as they wandered the world in search of food and a friendly climate. They didn't get to sit down with a leisurely cup of tea and ask about the mutual interests, family history, or favorite pastimes of the people wrapped in bear hide looming over them, spears in hand. They had to know quickly, almost instantaneously, whether or not they were safe. And if they weren't, they needed time to react. Knowing the answers to these questions within a millisecond or two gave them the time do so.

Although the circumstances of our lives have certainly changed as a result of modern society, our basic human experience remains much the same as it has always been. When we meet someone for the first time, we still need to know as quickly as possible 1) if we are safe and 2) if we can trust them. This is true in business, with friendships, and especially when it comes to love, romance, and intimacy. While scientists have traditionally attributed our first impressions of one another to brain functions, new evidence suggests that it's more than the brain alone making the judgment. The heart plays a vital role in helping us make split-second decisions.

The Institute of HeartMath, often abbreviated as IHM, is a pioneering research organization dedicated to exploring and understanding the full potential of the human heart, sometimes going beyond what is typically done in university laboratories and classrooms. I want to clarify that while I'm not an employee of IHM, for over 20 years I've worked with them closely to share many of their science-based discoveries with mainstream audiences.[29] Throughout the remainder of this book I will reference IHM research, discoveries, and techniques, with IHM permission, to illustrate the applications of embracing the heart's potential in our lives. A summary of studies conducted by IHM regarding intuition, for example, beautifully states the heart's role in our decisions:

> At the center of this ability [intuition] is the human heart, which encompasses a degree of intelligence whose sophistication and vastness we are continuing to understand and explore. We now know this intelligence may be cultivated to our advantage in many ways.[30]

As mentioned previously, it's because heart intelligence bypasses the filters of the brain (thoughts related to past experiences, self-esteem, and so forth) that it can make its decisions regarding our safety and well-being almost instantaneously. Alex Todorov's study found that we make a judgment in as little as a tenth of a second when we encounter a new face.

A number of additional studies have found that, just as our mothers told us in nonscientific terms, first impressions are generally spot-on. Because we live in a society that has traditionally discounted intuition in the past, however, we often find ourselves discounting our first impressions when it comes to the most important choices of our lives.

I have had friends, for example, who confided in me that the first time they encountered the person they would marry, their impression was to run away, and run fast! Rather than listening to their hearts' wisdom, however, they rationalized what they were feeling and did the opposite. From all outward appearances, there seemed to be no good reason not to go forward with their relationships.

In one instance, it was only after 12 years of marriage that my friend, a woman with whom I shared a corporate office, admitted to herself that her first impression of her husband had been correct. The man she'd married did not grow to respect her any more in the 12 years of their marriage than she felt he did when they first met. The key here is that she knew—her heart knew—almost instantly (in as little as one-tenth of a second) that the relationship wasn't safe. Because she ignored her heart's wisdom, she devoted 12 years of her life to arriving at the same conclusion. During those 12 years, she had experiences that empowered her to think differently about herself and accept that she was worthy of receiving more respect than her husband demonstrated.

When we hear of experiences like this, it becomes clear that rather than thinking in terms of black-and-white decisions that may look good on paper, we have the opportunity to be informed from a deeper wisdom that transcends the bias of the mind. Ultimately, it all comes down to our intuition and what we feel in our hearts.

AWAKENING OUR HEARTS' WISDOM

Embracing the benefits of our hearts' wisdom can immediately catapult us beyond traditional boundaries when it comes to the way we live, our capacity to solve problems, and even our capacity for love. It's these capabilities as well that give us the resilience to embrace big change in our lives—and to do so in a healthy way. When we take into consideration all that we now know about the heart—such as the fact that it is part of an extended neural network that was already developed when our ancestors appeared on earth 200,000 years ago; the fact that we have a little brain in our hearts made of cells that think, feel, and remember independently of the brain; and the fact that we can self-activate the benefits that come from the relationship of the brain and heart—the question now is, what else does the heart do that we're only beginning to understand? What capabilities await discovery today that we've either forgotten we possess or are only just now beginning to understand fully?

Key 19: The heart is the key to awakening deep intuition, subtle memories, and extraordinary abilities thought to be rare in the past, and to embracing these attributes as a normal part of everyday life.

chapter four

THE NEW HUMAN STORY

Life with a Purpose

"When we deny our stories, they define us. When we own our stories, we get to write a brave new ending."

— BRENÉ BROWN (1965–), AMERICAN RESEARCHER

When we answer the question *Who are we?* from the point of view of conventional science, is it possible that not only are we on the wrong track, we're stuck on that track, and it is leading us farther and farther from understanding the most empowering truths of our lives? Being stuck on the wrong track has happened before, and the scientific community is still reeling after discovering how far off their expectations were the last time an accepted theory was proven mistaken.

THIS IS NOT WHAT THEY EXPECTED!

At the completion of the Human Genome Project (HGP) in 2001, scientists were astonished to learn that the genetic blueprint for a human is about 75 percent smaller than they'd thought it would be. This wasn't just a little mistake in calculation. It was such a huge discrepancy from the original thinking that the international community of biologists and geneticists involved in the project had to acknowledge a difficult fact regarding their fundamental assumptions.

Before the HGP, it was believed that there would be a unique gene for creating each one of the unique proteins that make up our bodies. Based upon this idea of one-to-one correspondence, researchers expected that the project would identify at least 100,000 genes in the human genetic blueprint. Scientists and entrepreneurs were so certain of this, in fact, that they had planned to develop pharmaceutical products to modify and "fix" the genes that were discovered, and build an entire new industry of gene medicine, once the results of the project were known.[1] No one anticipated the actual results of this project. And when those results came in, scientists in universities, research institutions, and medical laboratories throughout the world had to come to terms with a surprising new reality.

The HGP revealed that there are only about 20,000 to 24,000 genes in the human genome, 75,000 fewer than had been expected![2] The question was, where were the "missing" genes? Did they even exist?

Further research done following the HGP revealed where the original thinking of the scientists was flawed. Rather than one gene coding for one protein, we now know that a single gene can produce the codes for multiple proteins, sometimes numbering in the thousands. One gene from a fruit fly, for example, can code for many as many as 38,000 different proteins.[3] The same principle appears to be true for humans, only to a lesser degree. "It seems to be a matter of five to six proteins, on average, from one gene," says Victor A. McKusick, co-author of the landmark paper that described the HGP findings in 2001.[4]

But how could such a fundamental error in thinking have gone undetected for so long? How could the basic assumption at the foundation of a futuristic new field of science, one that was believed would lead to creating an entire new industry of medicines, be so flawed?

The answer to this question is the reason I'm describing this account. *The mistake was due to the scientific acceptance of an unproven theory*—the assumption of the one-to-one correspondence between genes and proteins—that scientists had made years earlier in the mid-20th century.

Craig Venter, the president of a firm leading one of the HGP gene-mapping teams, recognized the significance of the HGP results immediately when he stated, "We have only 300 unique genes in the human that are not in the mouse. This tells me genes can't possibly explain all of what makes us what we are."[5]

The HGP illustrates a perfect example of the consequences of embracing a scientific assumption as fact in the absence of evidence to support it. In this instance, an entire field of science and medicine, and the people and industries relying upon the science and medicine, was thrown into chaos by the errors in judgment. The outcome of the HGP also forced the rethinking of a basic premise that had been wholeheartedly embraced by scientists and taught as fact in university classrooms. And while scientists now appear to be on the right track when it comes to the way genes and proteins are related, the Human Genome Project is not the only time an unproven doctrine has led scientists to a dead end in their assumptions. If it were, we could call what happened with the project an anomaly. But it's not an anomaly. The example of the HGP illustrates a way of thinking that we've seen before in the not-so-distant past.

SAME EXPERIMENT, NEW EQUIPMENT, AND A NEW RESULT!

The scientific belief that everything we can see and touch is separate from everything else is another example of the kind of thinking that has led to a scientific dead end. The idea of separation has its roots in the famous Michelson-Morley experiment originally performed in 1887. Named for the two scientists who designed the experiment, Albert Michelson and Edward Morley, this experiment was the much-anticipated effort of the scientific community to settle, once and for all, the question of whether or not a universal field of energy connects all things.[6] The thinking at the time was that if such a field actually existed, it would move in relation to the earth. And because the field would be in motion, it would be possible to detect that motion.

The experiment was performed in a makeshift laboratory in the basement of a building at Case Western Reserve University. The results of the experiment, as the data were interpreted by scientists of the time, were thought to prove that no universal energy field exists, with the implication that everything is separate from everything else—meaning what happens in one place has little effect, if any, on what happens somewhere else.

These findings became the foundation of scientific theory and classroom instruction for nearly a full century. Until Michelson and Morley's 19th-century experiment was repeated in the 20th century, multiple generations grew up believing that we live in a world where we're separate from one another and the world around us, and that what we do in one place has no effect beyond that place. This belief was reflected throughout our civilization in ways that ranged from personal choices that affected other people, and the growth of economic systems that benefitted some people at the expense of others, to the bigger picture of humanity's relationship to the earth itself. For scientists throughout the world, the assumptions of Michelson and Morley were accepted as fact . . . that is, until the experiment was revisited 99 years later.

In 1986, a scientist named E. W. Silvertooth duplicated the Michelson-Morley experiment in a study sponsored by the U.S. Air Force. Under the unassuming title "Special Relativity," the scientific journal *Nature* published the results. Using detection equipment that was much more sensitive than what Michelson and Morley had in 1887, *Silvertooth did detect the field, and it was moving just as Michelson and Morley had predicted it would 100 years before.*[7] In the process, he debunked a whole worldview.

For nearly a century, the best science of the modern world was based upon an idea that simply wasn't true. Fortunately we know better now and can apply this knowledge. Yet even with the later experiment proving the existence of the field and the vital role it plays in our lives, the principle of separation continues to be embraced today in some textbooks and taught in some university classrooms. Because of this, members of yet another generation are being led astray.

I'm identifying the Michelson-Morley experiment and the Human Genome Project as classic examples of how a scientific theory that's held in high esteem at one point in time can, and must, change when a new discovery overturns earlier assumptions. It's precisely this kind of discovery that is imploding the theory of human evolution, and it's of vital importance that our past assumptions must be personally abandoned and publicly dismissed when it comes to the belief that the DNA that makes us what and who we are formed purely by chance.

Key 20: Willingness to embrace a scientific assumption as fact, in the absence of evidence to support it, can lead us, and has led us in the past, to wrong conclusions when it comes to the way we think of ourselves and our relationship to the world.

IMPOSSIBLE ODDS

The conventional story of life on earth—the theory of evolution—asks us to believe that long ago, just the right conditions appeared in just the right way and at just the right time to create just the right environment for the right forces to form perfect atoms and forge them into the elements that gave birth to the first molecule of life. As if asking us to believe this unlikely series of events isn't already a stretch, we're then asked to further accept that this first molecule of life survived, and flourished, multiplying and diversifying countless times, and then triumphed through the ages via an adaptive strategy known as "survival of the strongest" to become the bodies that enable us to lead the lives that we lead today.

The odds that this series of events actually occurred are so small that they appear to be impossible.

The late two-time Nobel Prize–winning chemist Ilya Prigogine agreed. "The statistical probability that organic structures and the most precisely harmonized reactions that typify living organisms would be generated by accident," he said, "is zero."[8] In agreement with Prigogine, many other scientists, using the most advanced scientific methods available, are now able to tell us just how extraordinarily unlikely the chance origin of our DNA is.

Before his death in 1989, Swiss mathematician and physicist Marcel Golay calculated that the probability for even the simplest living protein to form by chance is 1 in 10^{450}, while plant physiologist and former head of Utah State University Frank Salisbury calculated the probability for the existence of a common DNA molecule as 1 in 10^{600}.[9]

These numbers are so unimaginably long and represent such a small chance of something occurring that I'm going to elaborate briefly here to illustrate just what the mathematicians are telling us. To clarify, the number 10^{600} is shorthand for the British unit of *one centillion* or 1 with 600 zeros after it. If we convert this notation into longhand and type it out, here's what it looks like.

1,000, 000, 000, 000, 000, 000, 000, 000, 000, 000, 000, 000,
000, 000, 000, 000, 000, 000, 000, 000, 000, 000, 000, 000, 000,
000, 000, 000, 000, 000, 000, 000, 000, 000, 000, 000, 000, 000,
000, 000, 000, 000, 000, 000, 000, 000, 000, 000, 000, 000, 000,
000, 000, 000, 000, 000, 000, 000, 000, 000, 000, 000, 000, 000,

000, 000, 000, 000, 000, 000, 000, 000, 000, 000, 000, 000, 000,
000, 000, 000, 000, 000, 000, 000, 000, 000, 000, 000, 000, 000,
000, 000, 000, 000, 000, 000, 000, 000, 000, 000, 000, 000, 000,
000, 000, 000, 000, 000, 000, 000, 000, 000, 000, 000, 000, 000,
000, 000, 000, 000, 000, 000, 000, 000, 000, 000, 000, 000, 000,
000, 000, 000, 000, 000, 000, 000, 000, 000, 000, 000, 000, 000,
000, 000, 000, 000, 000, 000, 000, 000, 000, 000, 000, 000, 000,
000, 000, 000, 000, 000, 000, 000, 000, 000, 000, 000, 000, 000,
000, 000, 000, 000, 000, 000, 000, 000, 000, 000, 000, 000, 000,
000, 000, 000, 000, 000, 000, 000, 000, 000, 000, 000, 000, 000,
000, 000, 000, 000, 000, 000

This huge number is the longhand version showing the odds that the first DNA molecule formed by chance. I'm emphasizing this point because scientists generally accept that when the odds of a possibility occurring are 1 in 10^{110} or more, the chances of that event happening are so small that it's impossible. If these numbers represented the chance of us winning the Powerball lottery, for example, we'd probably throw our tickets away because the odds would be so staggeringly small. So the scientists themselves are telling us that the fact that DNA exists at all represents a probability that's already "impossible" at 1 in 10^{110}, and this impossibility can be further multiplied by a factor of 5, to 1 in 10^{600}, to make it even more improbable!

English astronomer Sir Fred Hoyle and astrobiologist and mathematician Chandra Wickramasinghe placed the odds even lower in a book they co-wrote, at less than 1 in $10^{40,000}$, based upon the number of known enzymes needed for life and the chances of them appearing at random.[10] When we begin talking about odds this small, the numbers themselves become almost meaningless.

For the non-mathematician, Hoyle aptly described these outrageous statistics as being the equivalent of a tornado sweeping through a junkyard and assembling a Boeing 747 jetliner from scattered debris.[11] And it's through the eyes of this improbability that scientists are attempting to make sense of the origin of life. But if the evidence shows that we are the result of something more than the pure chance proposed by the theory of evolution, then the fact of our existence must take on new meaning as well.

Key 21: Renowned scientists tell us that it is mathematically impossible for the genetic code of life to have emerged through the process of evolution alone.

EVOLUTION: AN IMPOSSIBLY SQUARE PEG IN AN IMPOSSIBLY ROUND HOLE

When Darwin introduced his theory of evolution in the mid-19th century, it was believed that in the decades that would follow, new discoveries would further validate the theory that was already accepted as scientific fact in his day. What has happened since that time, however, defies this expectation. The evidence does not support human evolution. But rather than allowing the evidence to lead us to write a new story of human origins, there has been a concerted effort instead to force new discoveries into the framework of the existing story of evolution.

To recap, we see evidence of this in the efforts of mainstream scientists to draw a link between the ancient primate fossils of the past and modern humans on the primate evolutionary tree. With some mainstream media outlets, such as PBS, neglecting to offer their audiences a balanced perspective, as they did in their biased documentary on evolution, and with some academics, such as biologist Richard Dawkins, going so far as to demean and ridicule anyone who questions the conventional wisdom when it comes to human origins, the insistence that the evidence sustain existing theories is akin to the proverbial forcing of the square peg into a round hole. While it's certainly possible to pound a square peg until it's jammed into a round opening, it will never fit well because it simply doesn't belong there.

Discoveries about human DNA are telling us that our species doesn't fit into the neat and tidy traditional story of evolution. Nonetheless, people continue attempting to jam the facts into the theory in a way that's leading us away from properly solving the mystery of our existence.

OUR POINT OF NO RETURN

A friend of mine had a desktop computer that had been state-of-the-art, with all of the latest software, when she first purchased it nine years earlier. But as new updates for the operating system became available over time, such as enhanced network security, faster operating speeds, and system upgrades, she neglected to download them onto her computer. She was busy meeting work deadlines and didn't feel that the *New upgrades available* messages that showed up on her screen from time to time were a priority in her schedule.

During the first couple of years or so, the failure to keep her system up to date influenced her computer only in subtle ways. Some of the upgrades were small and affected few of her everyday computing needs. These little upgrades were noted as the number following the version: v1.1, v1.2, v1.3, and so on. But when the developers made big changes in the software that warranted an entirely new version, a v2.0, for example, it was a different story, because any new software began looking in her computer for the features in the previous version that it could build upon.

One day my friend was deep into the editing of a new book and she tried to open a file that she'd received from her editor, who was using another computer operating system. That's when everything changed and I received a phone call from her asking for help. "My computer is stuck! It's frozen and I can't even turn it off," she said.

After a couple of useless suggestions from me, it dawned on me—knowing my friend's aversion to software upgrades—what might be happening. "What version of your operating system are you using?" I asked.

Her answer told me the reason for her computer problem. My friend's computer software was literally years out of date. The software needed for her to read her book edits was relying upon the features of a recent upgrade to do its job—features that didn't exist anywhere on her system.

For my friend, the options were simple. She could either spend the afternoon downloading and installing each previous version of the software, one by one, to incorporate all of the updates she'd missed over time, or she could buy a long-overdue new computer that was completely up to date, with all of the latest software. Knowing the way my friend thinks about sustainability and maintaining her electronics, I was not surprised by her choice. She opted to spend the day updating her trusty old computer.

A NEW STORY ON AN OLD FOUNDATION

The story of my friend and her outdated software is an analogy for what the scientific community is experiencing today when it comes to expanding the theories of human evolution. When Darwin introduced his theory in 1859, it was a version 1.0 way of thinking. As new technology became available over time, helping science make incredible new discoveries about molecular biology and the human genome, the theory should have been upgraded to v1.1, v1.2, and so on.

But it wasn't.

The scientific method is based upon the principle of observation in individual research studies leading to "upgrades" in our common base of knowledge. Science is designed to be constantly updated and revised as new information comes to light.

What's happened, however, is that the reluctance—even the outright resistance—of the academic and scientific communities to acknowledge new discoveries related to human development over the past 150 years is akin to my friend's reluctance to incorporate occasional upgrades into her computing system. Now, seemingly all of a sudden, discoveries such as the DNA fusion on human chromosome 2 are changing the whole story. Trying to incorporate these types of discoveries into the existing story of evolution is like trying to download an entirely new version of software onto a computer that can't support it. The v2.0 DNA discoveries are so different from the original concept of evolution that there's no place for them. The v1.0 theory just doesn't fit the facts.

My friend attempted to do just what the scientific community is attempting to do today: to make an effort to somehow "fix" the existing software system on her computer so that it would accommodate additions. My friend discovered, however, that there's a point of no return when it comes to computers and the software they can support. The software that's written for a computer is directly linked to the nuts and bolts of the machinery: the chips, processors, and capacities they're built for. When advanced software begins to ask for volumes of memory or processing speeds that are not supported by the existing hardware, the new software can't be used. Despite my friend's best efforts to catch up to the level of the upgrades being offered to her, she was forced to invest in a new computer whose hardware could accommodate the latest versions of the software she needed to do her job.

This is precisely where we are when it comes to the story of human origins. The attempt to incorporate the story of exact, rapid DNA mutations, such as the ones we find on FOXP2 and human chromosome 2, into the existing story of the long, slow, and gradual process of evolution isn't working. And it can't, because the discoveries that preceded it have been discounted in evolution theory. We've reached the point of no return.

Just as my friend's trusty old computer, which had served her so well, reached a point where it was obsolete, we've reached a point where the human story we've told ourselves in the past is now obsolete. Now it's time for us to invest in a new theory that embraces the anomalous information scientists of the past couldn't account for.

Just as geneticists and biologists have had to shift their thinking to accommodate the evidence from the Human Genome Project, and just as physicists have had to update their theories to adapt to the most recent results of the Michelson-Morley experiment, we must make room for additional discoveries in the future that may upset some of the most cherished beliefs of our current leading thinkers. In a beautiful and perhaps unintentional way, it seems that science has already given us everything we need to do just that. The building blocks for the human story v2.0 already exist. It's all about how we choose to embrace what the evidence has already revealed.

AN UPGRADE FOR THE HUMAN STORY

In a way that is similar to the outcome of the Human Genome Project, the very science that was expected to eventually support Darwin's theory of evolution and solve the mystery of our origin has now done just the opposite. New discoveries are presenting unsettling implications that fly in the face of longstanding scientific tradition. Ironically, the evidence is leading us in a direction that now parallels some of our most ancient and cherished traditions about our beginning. For convenience, I'm including a condensed summary of the evidence described in the previous chapters here as the building blocks for the new human story.

Fact 1: The relationships shown on the conventional human evolutionary tree are speculative connections only. While they are believed to exist and are taught as factual in public schools, a 150-year search has failed to produce the physical evidence that confirms the relationships depicted on the evolutionary family tree.

Fact 2: If the fossil record is accurate, anatomically modern humans appeared on earth suddenly approximately 200,000 years ago with advanced features that set them apart from every other form of life that had already developed to date or and has developed since. These features have remained unchanged in us and include:

- A brain 50 percent larger than that of our nearest primate relative, the chimpanzee.
- Upright posture and advanced manual dexterity.
- The capacity for advanced language.
- An extended neural network that allows for extraordinary capabilities, such as deep intuition and access to heart-based wisdom on demand.

Fact 3: The lack of common DNA between AMHs and Neanderthals tells us that we anatomically modern humans did not descend from ancient Neanderthals. Additional studies reveal that our forbearers shared the earth with the Neanderthals that were previously thought to be some of our ancestors. Logically, if we shared the earth *with* them, we could not have descended *from* them.

Fact 4: DNA analysis reveals that:

- The DNA that sets us apart from other primates is the result of a mysterious process of "fusion" that resulted in the second-largest chromosome in the human body: human chromosome 2.
- The way human chromosome 2 was fused suggests something *beyond* evolution has made our humanness possible: the "turning off" or removal of overlapping functions, and the fact that it happened quickly, rather than slowly over a long period of time.

Armed with these four facts alone, we have more than enough reasons to rethink the traditional story of who we are. Clearly we're not the product of an evolutionary process, at least not the kind of evolution that Charles Darwin had in mind when he proposed his original theory in the 19th century. Looking at the scientific probability that the DNA that gives us our humanness occurred by chance, the odds of which have been compared to the odds of a junkyard tornado creating an airplane, points to the conclusion that we humans are not the result of random events set into motion by pure chance.

The question now is simply this: Are we willing to embrace what the best science of our era is showing us? If we answer yes, then we must also embrace a new human story that better reflects the evidence we've compiled. And while modern science is struggling with what the new evidence means and how it fits into the story of our origin, the indigenous people of the earth and practitioners of some of the most widely accepted spiritual traditions of the world are not. In their way of thinking, the modern evidence simply reconfirms and deepens their acceptance of the ancient accounts that are at the core of their beliefs.

With over half of the world professing to follow one of the three major religions that stem from a common history—Judaism, Christianity, and Islam—it's no surprise that the new scientific evidence is so well received by so much of the world.

ANCIENT ACCOUNTS OF AN INTENTIONAL ORIGIN

Almost universally, scriptures from the world's most ancient and cherished spiritual traditions agree that we humans are linked to something beyond ourselves and our immediate surroundings. And as different as these traditions are from one another, when it comes to the story of human origins their accounts are also surprisingly similar. Common themes include:

- An advanced intelligence and an intentional act being responsible for our origin.
  - o The use of the terms *they* or *angels* (in the ancient tongues the authors spoke) when describing human creation suggest that a group intelligence was involved.
- Descriptions explaining that we are the product of the dust/clay/soil of our planet being fused with an essence that is not from this world.
  - o In the three Abrahamic traditions, Judaism, Christianity, and Islam, it is the dust or clay of the earth that is used to create the first body of a human being.
  - o Following the formation of the first human body, for example, life is "breathed" into the nostrils of the person and the blood of a being of higher intelligence is mixed with the first person's body.

In great detail, the ancient traditions took extra care to describe the intimate nature of our creation and how we, like our first ancestors, are infused with what has been described as a special spark of a mysterious essence, eternally joining us with one another and with something we can't see that exists beyond our physical world.

While these details have been largely edited out of contemporary versions of the Christian Bible, ancient Hebrew literature, such as the Haggadah and certain "lost" scrolls, shows that this level of detail was intended in the original texts. It's this mystical spark, which science so far has yet to find ways to measure, that sets us apart from all other forms of life on earth.

The following are a few key examples of ancient stories that illustrate the common elements of story I am referencing.

The Sumerian creation story. The region that is now the country of Iraq was the site of ancient Sumer, traditionally thought to be the oldest civilization on earth. (New discoveries at sites of other early civilizations, such as Turkey's Göbekli Tepe, show that these sites may prove to be equally as old or older.) The Sumerian creation story was recorded on a stone tablet found in southeastern Iraq, in what was the ancient city of Nippur.

According to the creation story, known by archaeologists as the Eridu Genesis, Nippur is where the first human was created. The story describes a time when multiple gods ruled over the earth. For reasons that are detailed in the text, one of the gods was sacrificed, and his blood was mixed with clay to create the first human. An excerpt tells the story:

> In clay, god and man
> Shall be bound,
> To a unity brought together;
> So that to the end of days
> The Flesh and the Soul
> Which in a god have ripened—
> That soul in a blood-kinship be bound.[12]

In other words, this story suggests that we are the product of an intentional act that was overseen by advanced human-like beings, imbuing all of us with certain qualities that the gods placed into the new human.

The first human in the Jewish, Christian, and Islamic traditions. Among the recurring themes of ancient creation stories are descriptions of human origin at the hands of more advanced and otherworldly beings. The oral traditions of the Hebrew Midrash and early Kabbalah, for example, describe how the creator asked his angels:

> *Go and fetch me dust from the four corners of the earth, and I will create man therewith.*[13]

In similar terms, the holy Quran refers to God's creation of humankind from natural elements:

> *We created you out of dust.*[14]

At another point in the Quran, however, the birth of man is attributed to God, acting through fluid.

It is He [God] who has created man from water.[15]

While these last two descriptions may appear to be in conflict, a closer look at the verses resolves the mystery. In the first description, the story of Adam originating from dust is part of a larger sequence describing the events that led to the first living beings. The verses reveal that after Adam's origin as dust, there was a process of creating progressively more lifelike forms as the first human began to take shape. The description states that after the dust, the human was formed from

a small life-germ, then from a clot, then from a lump of flesh, complete in make and incomplete, that We may make clear to you.[16]

In this way, the Quran adds to the traditional descriptions of Adam's creation by filling in details of how "dust" becomes flesh.

In a similar way, in the Western world, when we ask someone what the first human on earth was made of, the reply is generally that we're made of the same "stuff" that the world is made of: clay, mud, or dust. To support such statements, we are often directed to the biblical creation story in the book of Genesis. Shared by nearly two billion people of the Jewish and Christian traditions, the story of Adam provides the most basic description of human origin. Deceptively simple in its form, Genesis recounts the miracle of human creation through a very few simple words.

The Lord God formed man from the dust of the earth.[17]

The Mayan creation story. From approximately 250 C.E. until 900 C.E., the Mayan civilization flourished across a vast area of North America, extending from what is now northern Mexico to the south, encompassing the entire Yucatán Peninsula and what are now the countries of Belize and Guatemala, as well as portions of Honduras and El Salvador. The Mayan civilization is recognized as one of the six "cradles of civilization" that appear to have developed in different places on the earth, at different times, independently of one another. The remaining five are Mesopotamia, and the civilizations of the Nile River, the Indus River, the Yellow River, and the central Andes of Peru.[18]

The ancient Mayans had a complex system of mathematics and hieroglyphic writing, an advanced knowledge of cosmic cycles, and a well-developed creation story. It's known today as the *Popol Vuh,* and it describes the theme of human creation in a way that is very similar to the story told in some of the original Semitic scriptures. The *Popol Vuh* tells us that the first attempt at human creation was flawed. Subsequent attempts led to a refinement of the creation process.

The point I'm making here is that the Mayans, with an advanced knowledge of the cosmos (which was confirmed only in the mid-20th century) attributed their existence to a conscious process invoked by an already-existing intelligence rather than a spontaneous and random process of nature. The *Popol Vuh* description begins:

> *Together they made a body, but it wasn't right. . . . We must try again.*[19]

The previous examples are only a sampling of elements common to many ancient and indigenous accounts of human origins. Although these accounts vary in their specifics, the general themes are remarkably consistent. They tell us that we are:

1. The product of an intentional act
2. As such, related to the greater existence of a cosmic family
3. Imbued with the traits of our creator(s)

These are precisely the themes that the theory of evolution, in its current form, cannot account for.

Key 22: Almost universally, ancient and indigenous traditions attribute our origins to the result of a conscious and intentional act.

EVOLUTION? CREATIONISM? OR . . . ?

The thinking of the past has been binary when it comes to the question of our origins. If evolution is not our story, the go-to alternative has automatically been assumed to be the story told by creationists of a divine beginning similar to the biblical account. With this kind of thinking, all the "baggage" of religious doctrine from the creationism side of the issue, and all the "baggage" of the science zealots who are clinging to fundamentalist evolutionary theory, has made it nearly impossible to explore a third possibility. Nevertheless, DNA studies are telling us that a third possibility exists.

The scientific fact of the mutation that made our FOXP2 gene and complex speech possible, and the DNA fusion that created human chromosome 2 and enabled the advanced brain functions associated with it, as well as the evidence that suggests these mutations cannot be attributed to evolution alone, all invite us to consider something beyond creationism and evolution when it comes to our species' origin. For the purposes of this discussion, and to honor the fact that mutations did occur, while acknowledging that something more than evolution contributed to the mutations, let's call our third possibility *directed mutation*.

The name says it all. Some force that is not presently accounted for in the scientific story is responsible for the precision, timing, and refining of the mutations that makes us who we are. That unknown force directed the mutations that science has now proven to exist. And while the phrase *directed mutation* is accurate in terms of what it describes, it also opens the door to the obvious question of who, or what, did the directing.

Of course, to even consider the possibility of directed mutation leads us into a realm historically reserved for religious explanations of our existence, or more recently, to extraterrestrial explanations that are beyond the purview of science—at least as we know science today. Because science is based in understanding nature and the many expressions of the natural world, a supernatural explanation of human origins by definition must lie beyond nature and scientific understanding.

My sense, as a scientist, is that the possibility of directed mutation is both beyond Darwin's theory and beyond creationism. Rather than requiring a supernatural explanation, the evidence, I believe, is leading us directly to a new and expanded understanding of the natural world and nature itself. This new understanding seems to hold the potential to catapult us light-years beyond the limited views of how we came to be that we've embraced in the past. In other words, through our willingness to embrace the deepest truths of our origins we may, at last, unlock the deepest mysteries of the cosmos and know our place in it.

This path of inquiry leads to what some scientists have called a Pandora's box of possibilities—once the box has been opened, the contents cannot be stuffed back inside. From the mystery of what makes us human, beyond the small number of genes discovered by the Human Genome Project, to the mystery of the mutations that resulted in the FOXP2 gene and human chromosome 2, the new human story is leading us to embrace an explanation for how our forbearers came to be anatomically modern—built just like us—that's beyond the pure chance of lucky genes and random mutations.

Our willingness to embrace the third option of directed mutation puts us squarely into the realm of unmeasured fields and unseen forces, and an unseen intelligence that science has been reluctant to consider in the past. And this is where the sea change begins when it comes to the scientific answer to *Who are we?*

When we allow for new interpretations of the existing evidence, the new conclusions that emerge can only serve to empower us with new possibilities when it comes to the way we think about ourselves and our potential. It can also give us new perspectives on the way we live our lives and solve our problems. Perhaps most importantly, it has the potential to change our sense of self-worth and our appreciation for the value of all human life.

Just as we devote hours of time today to searching dusty archives and genealogy websites to tell us about our family's past so we can understand ourselves as individuals, I believe that we also long to connect with the deeper truth of where we've come from as humans. We feel a greater sense of belonging and often a sense of pride when we explore our lineages and learn the things our ancestors accomplished and overcame to make our lives possible today. And that same sense of pride and belonging arises when we discover that our lives are the result of a conscious act of directed mutation.

I've spoken with biologists, anthropologists, and others in the scientific community regarding precisely the evidence that I've shared in the previous chapters and its implications. Their response has become predictable. At first, when they hear my suggestion that evolution is not our scientific origin story, they think I'm joking. Later, when they realize that I'm completely serious about what I'm suggesting, the tone of the conversation and the expression on their faces changes. Some people become aggressive and indignant. They take the possibility personally and ask why, as their friend, I would work to undermine their decades of teaching and their reputations.

Others, often in the same conversation, become quiet and withdraw. In private, they sometimes tell me that they've known this conversation was coming; they just didn't know when. It had to come, they tell me, because discoveries that were once classified as anomalies have continued to accumulate so quickly that it's clear science took the wrong path when it comes to solving the mystery of our origin. At the foundation of the newly emerging story of human life is another story that is unfolding, on the grand scale of the universe itself, that describes a different kind of life.

THINKING "DEAD" IN A LIVING UNIVERSE

For over 300 years, the scientific story of the origin of our universe has led us to believe that we live in a "dead" universe. From this perspective, the cosmos is made of inert stuff, like the dust of exploded stars or debris from colliding asteroids and disintegrated planets. In a dead universe, there is no point to life and no reason for living. But new discoveries by leading-edge researchers are giving us very good reasons to rethink the dead universe story, which means there may be a purpose to life after all.

At the forefront of defining how the new scientific paradigm of a living universe may affect us in our daily lives is social researcher Duane Elgin. Elgin's philosophy, based upon existing evidence in the scientific community, accepts that the universe is a living entity that's growing and evolving rather than a lifeless system. He shows us that the way we think of the universe and our place in it is at the very foundation of the way we live our lives and solve our problems, especially how we treat one another.

If it were true that we live in a dead universe, then it would actually make sense to do what we've already done in the past, which is to exploit every resource available to the highest degree possible and reap the rewards of those resources. In Elgin's words, we relate to our belief that we're in a nonliving universe "by taking advantage of that which is dead on behalf of the living. Consumerism and exploitation are natural outcomes of a dead universe perspective."[20] This is how humanity has been living up to now, with rare exceptions.

It's no coincidence that Elgin's description of consumerism and exploitation reflects the world that we find ourselves in today. Just as the theory of evolution led us to believe that human life is the result of chance events, we've also been led to think of the universe as a resource that's ours to dominate and exploit.

The problem with this mind-set is that, ultimately, it has led to the depletion of natural resources, unsustainable forms of food production, and the conflicts over scarce resources that are at the root of so much suffering today.

But Elgin believes that we are part of a living system, and that knowing the truth will change how we relate to one another and lead us toward a more sustainable lifestyle of cooperation. The parallels that exist throughout the universe, in every known living system, lend credence to this view. From microbes and neural networks to ecosystems and the behavior of entire populations, all living systems, regardless of their size, show characteristics demonstrating the sharing of energy and information. In support of his theory, Elgin describes how the universe is:

- Completely unified and able to communicate with itself instantaneously in nonlocal ways that transcend the limits of the speed of light
- Sustained by the flow-through of an unimaginably vast amount of energy
- Free at its deepest, quantum levels[21]

While Elgin is quick to admit that these traits in and of themselves don't mean that we are part of a living universe, he notes that each fact adds to a growing body of information that supports this theory.[22] By extrapolation, as living beings we are part of this exchange of energy and information. Our existence has a purpose that is greater than paying our bills on time.

Key 23: A growing body of evidence suggests that we exist as part of a living and vibrant universe rather than one simply made of inert dust, gas, and empty space.

IN A LIVING UNIVERSE, LIFE HAS A PURPOSE

In a universe that's alive, it makes sense that living systems would appear often and in many ways. It makes sense because life itself is the force that's driving the system. To discover that we exist as living beings within the context of an even larger living system implies that our lives are about something more than simply being born, enjoying a few years on earth, and dying. It implies that somewhere, underlying everything we know and see, our lives have purpose.

And this is where our story takes us beyond the realm of proven science.

Key 24: If we're the result of something more than pure chance, then it makes sense that our lives are about more than purely surviving. It implies that our lives have purpose.

As a society, we now find ourselves at a meeting point of two ways of thinking about ourselves and the universe we live in. Elgin's living universe offers us the big picture of life having a purpose from the top down—from the macro scale of the universe itself as a living entity, within which, at the micro scale, the living cells that make up our bodies express themselves. The discoveries that I've shared in this book offer the evidence from the bottom up—from the micro world of mutated DNA yielding more complex expressions of life within the macro context of Elgin's living universe.

When we consider the universe as something that's alive, it changes everything. Elgin's words offer a beautiful sense of that perspective.

> In a living universe, our physical existence is permeated and sustained by an aliveness that is inseparable from the larger

> universe. Seeing ourselves as part of the unbroken fabric of creation awakens our sense of connection with, and compassion for, the totality of life. We recognize our bodies as precious, biodegradable vehicles for acquiring ever-deepening experiences of aliveness.[23]

Herein we may find our answer to the question of life's purpose. The existence of a living universe tells us that we are part of the world around us, and not separate from it, and that our aliveness is part of a greater aliveness. And as the very goal of life in the universe is to grow, change, and perpetuate itself, these are precisely the qualities that we should strive to embrace throughout the course of our time in this world as human beings.

Through each experience we face in life—through the satisfactions and the frustrations of every job, through the ecstasy and the heartbreak of each intimate relationship, through the unspeakable joy of bringing a child into this world or the unbearable pain of losing a child, through the choice to take another human life and the ability to save a life, through each war that we create and every time we end a war—in all of these experiences and so many more, we learn to know ourselves better as individuals and as a species.

On an unspoken, possibly subconscious level, we may be creating precisely these experiences in order to push ourselves to the very edge of what we believe is true about us and what's possible in life. And each time we push ourselves to our edge and grow, we discover that there's more to know. We get to experience our aliveness and relish it if we choose.

This is the very definition of a living universe and our role in it. Our lives and lifetimes are our way of infusing the essence of our unique experience into an already living and extremely diverse entity. Perhaps Ray Bradbury says it best:

> We are the miracle of force and matter making itself over into imagination and will. Incredible. The life force experimenting

with forms. You for one. Me for another. The universe has shouted itself alive. We are one of the shouts.[24]

Within the limits that science has placed upon itself today, there is no direct way to know the purpose of life with certainty. Indirectly, however, the answer to the question of life's purpose may be hidden in plain sight. We may discover that the very existence of our advanced capabilities—our intuition, sympathy, empathy, and compassion—holds the key to solving this mystery.

Albert Einstein's work in science led him to precisely this conclusion. As is the case with so many scientists who strive to unlock the deepest mysteries of our existence, the deeper their discoveries take them, the more they recognize that there's something more to human existence than a sterile and meaningless universe would produce by accident. When Einstein was asked about the meaning of our lives, his response was elegant. I've included a relatively long excerpt of Einstein's thoughts to give context to the answer that I've italicized.

> *A human being is a part of the whole, called by us the "Universe," a part limited in time and space. He experiences himself, his thoughts and feelings as something separate from the rest—a kind of optical delusion of his consciousness. This delusion is a kind of prison for us, restricting us to our personal desires and to affection for a few persons nearest to us. Our task must* be to free ourselves from this prison by widening our circle of compassion to embrace all living creatures and the whole of nature in its beauty. *Nobody is able to achieve this completely, but the striving for such achievement is in itself a part of the liberation, and a foundation for inner security* (author's emphasis).[25]

The beauty of Einstein's statement is that it transcends numbers, statistics, and logic. It's a purely intuitive answer to a serious scientific question. It's also a perfect example of how advances in modern science have carried us to the edge of what science can tell us with certainty. There's a place—an unspoken boundary—where the nuts and bolts of scientific explanation fail when it comes to describing life. They do so because we're more than cells, flesh, and bones. There's a quality to human life that simply cannot be defined in purely scientific terms, as we know science today. And it's that quality that potentially can lead us to comprehend the deepest truths of our existence.

For the scientific community to embrace the fact that evolution can no longer be our story, they say, would be like a wrecking ball appearing one day and leveling 150 years of exploration and hard work as well as lifetimes of teaching that the work has created. I can certainly see why some people would think this way. No one wants to see the foundation of his or her life's work demolished.

But I can also see something very different happening. As important as science is in the world today, as we push the boundary of scientific knowledge to the very edge of its capacity to define the world, we discover the limit of its capacity to serve us. And this is where science, as we know it today, breaks down. There are qualities of human life that simply cannot be measured and defined.

SCIENCE CAN'T MEASURE THE CAPACITY TO LOVE

In some respects we may think too highly of science. We may give too much credit to what we believe science can accomplish. Maybe we've put science and the scientific method on such a high pedestal that we simply assume science either already has the answers, or that it has the potential to solve the deepest mysteries of life, such as our individual life purpose. And if this is the case, it may be because we are asking too much of science when it comes to the question *Who are we?*

The German philosopher Karl Jaspers reminds us of this when he says, "The limits of science have always been the source of bitter disappointment when people expected something from science that it was not able to provide."[26]

The "bitter disappointment" that Jaspers is describing may be precisely the source of the frustration that we see in the scientific community when it comes to reconciling new discoveries with the existing theory of human origins. We may be asking science to do something that it cannot do and was never designed to do. I say this because of the nature of science itself. Science can only tell us *how* the molecules of our bodies behave now and how they have behaved in the past. But science can't tell us *why* those molecules appeared to begin with.

One of the reasons science is incapable of providing this answer is because scientific information is based upon events that are either observed in nature or duplicated in the laboratory to prove a theory. The fact is that no one living today witnessed the moment when the first human life appeared on earth. And in the laboratory, the process that would make such an awesome event possible has never been reproduced.

Although there are written accounts of human creation linked to religious traditions, made long after the fact, there is no first-hand record of the actual moment of human creation existing today—except for the creation itself: us. If we're going to solve the "why" of our origin in a living universe, we must look *beyond* the process of how we've arrived at our place today and instead consider what we've gained from our journey.

To do so may not be as difficult as it sounds. The clues that lead us to know if there's a purpose for living may be readily accessible within each of us, where they have always been. They live within each of us in the extraordinary abilities our genetic makeup gives us and in the way our expanded neural network of heart-brain communication empowers us.

Key 25: Our capacity for deep intuition, sympathy, empathy, compassion, and the self-healing that allows

> us to live long enough to share these capacities, are the needle of a compass that points us directly to our life purpose.

No other form of life on earth has the capacity to love selflessly, to embrace change by choice in a healthy way, to self-heal, to self-regulate longevity, or to activate the immune response on demand. And no other form of life has the capacity to experience deep intuition, sympathy, empathy, and ultimately, compassion, all of which are expressions of love—and to do so on demand. These uniquely human experiences are telling us that our lives have purpose, and the purpose is simply to embrace these abilities in order to know ourselves in their presence.

PART II

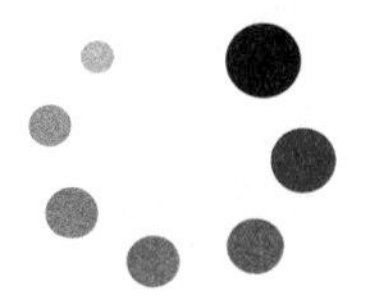

awakening the new human story

chapter five

We're "Wired" for Connection

Awakening Our Powers of Intuition, Empathy, and Compassion

"The only time we waste is the time we spend thinking we're alone."

— Mitch Albom (1958–), American author and journalist

Have you ever had one of those moments where you suddenly seem to be at one with the whole universe? One minute you're going about the mundane routines of everyday life, and an instant later, you unexpectedly find yourself in complete and total harmony with all life, all people, and the whole world. Maybe you were sitting in your car or truck at a stoplight, waiting for the light to change from red to green. Or maybe you were just staring out your car window as you waited to pick up your children after school.

Regardless of the scenario, it's generally when you're not focused on anything in particular that "it" happens. It's while you're *between* thoughts and not focused on anything specific that a feeling wells up from deep inside of you. Maybe your body floods with warmth. Maybe you get goose bumps on your arms or a prickle on the back of your neck. Then, suddenly, it's as if the veil between the worlds cracks wide open and you're given a front-row seat to viewing the meaning of your life, receiving answers to all your questions and clearly seeing the roadmap to fulfill your destiny.

And then, just as suddenly as it began, it ends. The light turns green. The driver behind you leans on the horn, urging you to cross the intersection. And *poof*! Just like that, the clarity of what you saw only seconds before evaporates. It's gone! And you're left to focus on the world of the guy honking behind you and what you'll make for dinner. You're also left wondering where your clear insight into the meaning of life has gone.

CONNECTED WITH EVERYTHING EVERYWHERE

While the previous scenario may be a bit of an exaggeration, it's probably not that much of one. We've all had moments of crystal clarity when we feel that we're "in the zone." We feel that we're just where we're supposed to be, at just the time that we're supposed to be there, in perfect harmony and attuned to the world. It can feel timeless when we're in the zone, because we're not thinking about anything. And that's the key. The moment we begin to analyze our experience, the zone collapses. It does so because when we think, we're jolted from the natural "go-to" place for our awareness when we're not thinking about anything—the heart—to the place where it takes an effort to maintain our focus—the mind.

The zone that gives us a feeling of connection, confidence, all-knowing, and peace is a natural state of being that we know as *intuition,* which begins in our hearts. Heart-based intuition bypasses the conventional reason and logic of the thinking brain. It taps into something that's deeper and more ancient than abstract reasoning, and yet, for most people, intuition has a familiar feeling. When we think about it, this familiarity should come as no surprise. Intuition is the internal language that our bodies have used to communicate with us since birth. We feel in our cells long before we ever learn to speak with our voices. And because we do, it makes perfect sense that this most primal form of communication—intuitive feeling—would be the language that our body uses to communicate vital messages regarding trust, safety, and survival.

The previous examples of times when we experience harmony and connectedness unintentionally, while doing nothing in particular, illustrate a specific kind of intuition: *spontaneous intuition*. This is the kind of intuition that comes when it wants to. It's also the kind that seems to leave when it wants to—usually before we're ready. The question is, can we trigger this powerful form of intuition intentionally when we need it the most? How can we trigger deep intuition on demand?

THE IMPULSE TO CONNECT

Sometimes our flashes of spontaneous intuition show up in simple ways, such as when we pick up the telephone to call a friend or loved one only to discover that they're already on the line that we were dialing. There was a time in my life when I experienced this kind of intuition with my mom. We had a standing ritual for a phone call every Sunday. From wherever I was in the world, I would do my best to call Mom, catch up on what had happened in her life during the week, and share with her what had happened in mine.

Following her divorce from my father in the mid-1960s, my mom chose to live alone. Our physical visits were few and far between and it was the weekly phone calls that kept us in touch.

There was a mystery that would often occur during those Sunday phone calls that illustrates the kind of intuition I'm describing in this chapter. I would pick up the receiver with the intention of dialing Mom's number, only to hear her voice already on the line without the phone ever even ringing.

"Hi!" she'd say. "This is Mom."

"I know," I'd answer. "I was just going to dial your number, but you're already here."

She was less surprised than I was by the whole thing, and more playful.

"See," she'd say. "We're just connected—psychic! Our ESP is working today!"

We'd laugh, and it was always a great way to begin our weekly catch up.

I'm sharing Mom's story here to illustrate a point. The connection between two people that makes such a simultaneous phone call possible, such as I had with my mom, is not the product of a conscious thought. It's not about making an appointment on a planner to call on a certain day at a certain time. In fact, it's almost impossible to create such a deep connection consciously. It's the thinking process of when and where to place such a call that creates the interference, preventing the intuitive connection from happening.

When I pick up the phone to call my mom, the moment that I do, it's in response to a subconscious cue. It's more of a feeling that it's the time to call, rather than the thought *Now is the time to make my call.* I'll be going about my typical daily routines and suddenly have an impulse—an intuitive hit or urge—to pick up the phone and make the call at precisely the instant that I do. And it's *because* I'm responding to an intuitive cue that it's not uncommon for my mom to be on the line already. The intuition that I feel to place the call is in response to her anticipation that we're about to connect. If I were to think about making the call, and actually do so even a second sooner or later, I'd lose the moment and never make the intuitive connection with my mom.

When it comes to intuitive experiences in our lives, almost immediately we discover two universal themes:

- The impulse to connect is generally not a conscious thought.
- The mutual impulse for connection appears spontaneously when we're not looking for it or expecting it.

IS IT INTUITION OR INSTINCT?

After we experience a deep intuitive connection in our lives such as the wisdom that appears at a stoplight or the phone connection I shared with my mom, and the experience ends, questions arise: Will it ever happen again? And if so, when? Do we simply wait for the universe to tap us on the shoulder, hoping that the next intuitive experience will be available for us when we need it, or is there more to it than that? Are we somehow empowered to open our intuitive connections when we want them?

These are good questions. And as different as they may appear from one another, the answer to each question is found in the same place. It's all about how we experience intuition for ourselves. The word *intuition* itself means different things to different people.

So let's begin at the beginning. What is intuition, and how does it show up in our lives?

Intuition is direct knowing that results from the way we receive sensual, conscious, and subconscious perceptions. As mentioned previously, the key here is that our intuition isn't based in reasoning. Rather, it's a subconscious assessment of the present moment based upon factors that may include street smarts, personal experience, and our physical senses, as well as instinct, in a way that gives us an awareness that is not thought out logically. Through our intuition, we can draw upon these factors and process them quickly without taking the time to actually think about them. This awareness is sometimes described as the soul's compass, because it helps us know what's right and true for us in a given moment. American author Dean Koontz describes this sense well by stating, "Intuition is seeing with the soul."[1]

There is a difference between experiencing intuition and the related phenomenon of instinct. *Instinct* is nature's way of informing us quickly of what's best for us and how to react in the present moment through responses that are "preset" or "hardwired" in our subconscious minds. Our instincts are based upon experiences of the past. And while the past that's informing us can sometimes be our own personal past, it can also include the collective past of our ancestors' response to a similar situation. When something has been experienced many times by many people, it becomes deeply ingrained in the collective psyche.

An example of this would be a child's innate fear of being left alone in a grocery store aisle, even if only for a few seconds, while the child's mother or father steps away to grab a can of soup. In the brief instant that children look around and realize their parent is gone, more often than not their response is predictable. It's common for them to cry in distress, or even to scream in terror at the realization that they're suddenly alone.

What makes this example so telling is that children may, in fact, be sensing very real dangers, even though they've never had a bad experience in the past to justify their fears. When something like this happens, there's a good chance that the child's fear is based upon instinct.

Our instinctive responses are based upon the collective experience of many people, over the course of many generations, who've learned, as in the preceding example, that it's safer to be with others in a familiar environment than to be alone in a strange one.

The child's fear is a shared primal instinct for safety and survival kicking in on a subconscious level.

Typically our instincts don't take into account the factors of personal knowledge and experience that can influence a subconscious response. Our instincts may tell us, for example, that we need to lash out and defend ourselves against friends or co-workers that we feel have attacked us through their criticism. Whether we're feeling threatened by the tip of a flint spear from an intruder in our cave 10,000 years ago or experiencing the "spear tip" of hurtful criticism from someone we know today, the instinct is the same—when we feel attacked, we react quickly and forcefully to defend ourselves. In the same situation, however, intuition may tell us that a softer, measured response is more appropriate.

Because our intuition is taking into account additional factors beyond our hardwired instincts, we can respond in a more thoughtful and less hurtful way. An example of this would be our personal history with the person criticizing us, the knowledge that they genuinely care about us, for example, and that what we perceived as a personal attack was really intended as constructive criticism. In such a scenario, the instinct to defend ourselves is still present, but we have the intuitive wisdom to temper our response. We can let the friend or co-worker know we feel attacked by their criticism without counterattacking in a hurtful way. Tailoring our response to the moment in this way has the potential to save us from doing irreparable damage to our relationship.

Key 26: *Intuition* is a real-time assessment that draws upon personal and past experience, sensory cues, and street smarts, while *instinct* is a response that is "hardwired" into our subconscious as a survival mechanism.

KNOWING THE DIFFERENCE

While we may not recall our own childhood response to being left alone, as adults we commonly find ourselves in situations where our instincts tell us that something isn't right and we may be in danger. The uneasy feeling we get while we're walking down a dark street in an unfamiliar neighborhood at 1 A.M. is a perfect example. Although we may have never personally had a bad experience walking down any dark street late at night anywhere in the world, other people have. In addition to the street smarts that may relate to a specific street or part of town, our instinctive fear is largely a subconscious response based upon the accumulated experiences of many humans walking in the same conditions of dark streets, late at night, for hundreds of generations.

Similar to the way it can be scary for a child to be alone in a strange place, dark streets often meant trouble for those walking them alone in the past, and we have the same concerns today. It's under the cover of night, when fewer people are on the streets, for example, that it's easier to be surprised by someone with bad intentions. When we're faced with walking alone down a dark street late at night, our instincts wake up to remind us of our collective experience and to prepare us for the possibility of a similar experience in the present moment.

I'm making the distinction between intuition and instinct here because of the way intuition happens. Rather than reacting exclusively from a library of past experiences, when our intuition kicks in, it informs us of what's true now, in the present moment. It can do so quickly, in real time, because it doesn't need to sift through the filters of all the experiences of dark streets in our collective past or the latest newspaper accounts of local crimes. Our intuition begins in our hearts—specifically with the little brain in our hearts. This is a collection of specialized cells that think, feel, and remember independently of the brains in our heads or our gut instincts.

Our intuitive responses and instincts can sometimes contradict one another, and it's easy to become confused when they're pointing us in different directions at the same time. While instinct may be telling us the darkness of the street scene isn't safe, the heart may be telling us that on this particular street, in this particular moment, we are safe. So what are we to do in such an experience? How do we know which voice is from the gut and which one is from the heart—and which one to follow?

While we all experience instinct and intuition on an almost daily basis, we discover our greatest levels of self-mastery when we can discern one from another and reconcile them in our lives. To do so we need a clear understanding of precisely where intuition comes from.

THE SINGLE EYE OF THE HEART

Part of my heritage is Cherokee, from America's Southeast. In the Cherokee language, there is a term for the intuition that already exists within each of us beyond logic and reason: *chante ishta,* which is pronounced "shawn-tay eesh-ta." Just as the Sanskrit word *prana* has no equivalent in the English language and must be translated loosely as "life force," *chante ishta* has no direct translation. An approximation of its meaning is "single eye of the heart" or "one eye of the heart."

Chante ishta is information that comes from the heart's natural wisdom. Another way of saying this is that intuition is a knowing made possible through the specialized cells that form the secondary brain in the heart. Our heart's cells are wired to sense the present moment and inform us about our immediate environment. And while our brains may listen in and respond to what our heart cells detect, they don't necessarily have to do so.

We have the ability to listen to the heart's wisdom independently of the brain and our instinctive and learned responses. The key is to avoid filtering the information that we're receiving from the heart through a subconscious library of instincts. The value of such wisdom is that it offers us a clear perspective on people's actions, life events, and situations that is beyond the polarities of judgment, beyond bias, and beyond fear.

THE WISE USE OF POWER

The heart has no knowledge of rules of social conduct or the legalities established by local and federal legislators. It doesn't know about the rightness or wrongness of culture, society, politics, and political correctness. The single eye of the heart knows only what's true for you in a given moment in time. It offers you a point of reference when there's no one to ask or turn to when you're faced with a difficult choice in your life. In doing so the wisdom of your heart offers you an unfiltered, uncensored, unbiased report of your immediate situation.

That being said, there is a responsibility that comes with anything that empowers us in life. When it comes to the power of heart wisdom, our responsibility is to use our power wisely, with common sense, in a way that is honoring to ourselves and kind to others. What I'm saying here is that the heart's intuition can be a useful guide in life, but it is not the basis for a life manual of rigid rules that we become enslaved by.

It's up to you to apply what your heart tells you wisely, balancing your intuition in a healthy and responsible way that makes sense under the circumstances of the moment.

THE SCIENCE OF INTUITION

Many of the recent discoveries regarding intuition and what it means in our lives have been made by scientists at the Institute of HeartMath. Similar to the conclusions reached by scientists early in the 20th century, what modern IHM studies suggest is that the function of our hearts is much deeper and much more subtle than has been previously recognized.

If we can understand the conditions in the body that support intuition, then we can re-create those conditions when we choose, rather than waiting for them to happen occasionally and randomly, as with my mom's phone calls. Fortunately, after two decades of investigation, researchers at IHM have developed the techniques to help us do precisely that. A study they performed in 2007 provided some of the first scientific evidence regarding what happens in our hearts and our brains during intuitive moments and suggests how we can re-create those conditions intentionally.

The purpose of the IHM study was to investigate one of the strongest emotional connections known to exist: the intuitive bond between mother and child. Based upon previous findings showing that "signals generated by the heart have the capacity to affect others around us,"[2] in this particular study monitors were used to measure both the mother's brain waves (as a conventional EEG) and the baby's heartbeat (as a conventional ECG, sometimes known as an EKG) while the mother held the baby in her lap. The prediction was that the interaction between the electrical fields of the baby's heart and the mother's brain would alert the woman to her baby's needs.[3]

At first, the influence of the baby's heartbeat was undetectable in the mother's brain. However, when the mother was asked to shift her attention and focus specifically upon her baby, the wave pattern in her brain changed in a profound and unexpected way. *When the mother focused her attention on her baby, the heartbeat of her baby was mirrored within her brainwaves.* The study concluded that the intentional act of shifting her awareness to her baby made the mother more sensitive and attuned to the electromagnetic signals of her baby's heart.[4]

While this study has meaning for many areas of our lives, the reason I'm sharing it here is best summarized in the words of the scientists themselves: "These findings have intriguing implications, suggesting that a mother in a psychophysiologically coherent state became more sensitive to the subtle electromagnetic information encoded in the electromagnetic signals of her infant."[5] *Coherence* can be defined as an energetic harmony that is established as an electrical signal between two organs in the body—in this case, between the mother's heart and her brain.

Continuing studies by IHM and other research institutions now suggest that the kind of intuitive connection demonstrated by the mother and her baby can be expanded to include our ability to attune our brain waves to the subtle energy fields of other people for reasons that range from emotionally supportive and healing prayer to informational connections, regardless of the distance between us and them.

Perhaps not surprisingly, the results of the study parallel what used to happen between my mother and me during our Sunday afternoon phone calls. They also help explain how a mother who has a son or daughter serving in a military combat zone half a world away could be attuned to what's happening in her child's life, like Kaye Young was to events occurring in the life of her son, Ronald.

REAL-WORLD INTUITION

In 2003, Ronald Young, Jr., was a chief warrant officer in the U.S. Army serving with the Fourth Brigade, First Cavalry Division, based out of Fort Hood, Texas. It was on a Sunday evening that his mother had a feeling—an intuitive sense—that her son was in trouble. At the time he was piloting an Apache helicopter on a military mission southwest of Baghdad in Iraq. In Kaye's words: "I just had a mother's feeling—I felt like Ron was there with me. I felt like he put his arms around me."[6]

Not long after Kaye's private premonition, her fears were confirmed. Military officials paid a visit to the family home and informed Kaye and other family members that Ron's helicopter had gone down the previous evening in the city of Karbala. Information was sparse, and Ron's whereabouts were unknown. He was listed by the Army as missing in action.

When she heard the official confirmation that Ron was missing, Kaye recalls that she immediately screamed, "I knew it! I knew it! I knew it!" And she did know it. While she may not have known the specifics of what happened, she knew—because her intuition had already informed her—that her son was in trouble. It was only through a television report shown in the United Arab Emirates capital of Abu Dhabi that the family learned of Ron's fate. He and another pilot were shown alive and in captivity. In the film they were talking with someone who could not be seen on the camera. And although they were prisoners of war, they appeared to be in reasonably good condition.[7]

Fortunately, there's a happy ending to this story. Ronald Young was freed from captivity in a daring rescue performed by U.S. Marines in April 2003. Rescued along with Young was the other pilot of his helicopter, David S. Williams, and five other POWs from the 507th Maintenance Company.[8] The intuitive connection that Kaye Young had with her son gave her insight into her son's experience before anything was known officially. It's a powerful example of how important information about our loved ones can appear spontaneously in our everyday lives.

Key 27: The emotional bond that exists between a mother and her children is now scientifically documented through studies that offer insights into the intuitive connection that we all can develop in our relationships.

INTUITION ON DEMAND

In the previous examples, the intuitive connection between people occurred spontaneously. They did nothing special in their lives to consciously initiate the experience. It just seemed to happen. It is common to experience this kind of intuition with people with whom we have strong emotional ties, because what happens in their lives is significant to our lives, as well. The technical name for this intuitive experience is *psychophysical coherence,* often shortened simply to *coherence.*

Now, although it can be a beautiful experience to have a deep connection with another person, when it's a spontaneous experience, it is difficult to rely on it to guide us in the moments that we need it the most, because we never know when or if the experience is going to happen again.

If we're just sitting somewhere waiting for the veils to drop and the universe to show us the right medical decision to make, the right job offer to accept, the best moment to end a relationship, or whether or not we should make a phone call to a friend we're concerned about, we may find ourselves waiting for a really long time. The reason is that spontaneous intuition is just that—spontaneous! It happens when it wants to, not always when we need it.

This is where intuition on demand comes in.

Just as it's possible to turn on our televisions and tune in to a box-office-hit movie in our own homes on the day and the time that we choose, we also can create coherence between our brains and hearts and trigger the deepest possible states of intuition when we choose. It's our ability to trigger our deep intuition consciously and intentionally that awakens the wisdom of the heart that may have formerly seemed sporadic and elusive in the past. When we think about the connection between Kaye and her son in Iraq, we begin to glimpse the untapped potential of such ability in our lives.

That potential is available for us in our everyday lives. And typically it's in the throes of what everyday life presents to us that we need our intuition the most. Deciding whether or not to go forward with leading a group into Egypt in the late 1990s is a perfect example of the kind of question that has no clear answer. It's also a perfect example of a time when the intuitive guidance of the heart was clear, direct, and accurate.

A LIFE-OR-DEATH DECISION MADE IN THE HEART

In November 1997, I was scheduled to take a tour group into Egypt. This was part of an annually scheduled pilgrimage I had led since 1992. To say that traveling Egypt is an amazing journey would be an understatement—it's beyond amazing! To actually stand in front of the Sphinx, a mysterious figure that I had studied in pictures as a child, or at the base of the Great Pyramid, looking up at over 400 feet of stones once covered in veneer, now naked and visible, was an experience of a lifetime. And I was under contract to lead a multinational group into the Egyptian desert to have precisely these sorts of experiences.

Then the national media began carrying horrific images of the events occurring on November 17 during the evening news. Though the details were still emerging, the gist of the story was clear. Armed terrorists had killed 58 foreign tourists and 4 Egyptians in a particularly savage attack at the temple of Queen Hatshepsut, a popular archaeological site near the city of Luxor.[9] My group was scheduled to leave for our tour, which included a stop at the place of these killings, the following week.

What has now become known as the Luxor Massacre was devastating for the country of Egypt on a number of levels. The tourism industry was crushed. Hundreds of tour companies immediately canceled their tours and pulled out of the country. Airlines stopped flying into Cairo. Hotels were empty. And the pride of the Egyptian people was suffering a deep hurt. "This is not us," my Egyptian friends told me on the telephone, pleading, "Please don't think of us in this way."

Immediately I began receiving phone calls regarding the planned tour. The people who were signed up for the trip begged me not to cancel. The Egyptian authorities were concerned about the possibility of another attack. And the tour company was waiting for me to make a decision and to do so quickly. Family and friends urged me not to go. The choices were clear: I could postpone the trip until another time, cancel it altogether, or go forward with the trip as planned. I felt pulled from all sides. Everyone I spoke with had an opinion, and their opinions all made perfect sense. And just when I thought I had made my choice, someone would call me and supply a good reason to make another choice. Clearly, this was one of those times when the decision was not black and white. There was no right or wrong in the situation and no way of knowing what would happen over the course of the following days and weeks. There was only me, my instincts, my intuition, and my promise to honor my group and myself with the best choice possible.

THE LANGUAGE OF THE HEART

Overwhelmed with the chaos of information and opinions, I turned off the telephone to shut out the input from other people. From my home in the high desert of northern New Mexico, I went for a long walk down a dirt road that I had walked many times in the past when I'd needed to make a tough decision. And I applied in my life precisely the technique that I will share later in this chapter to create coherence between my head and my heart, to get in touch with my deepest intuition regarding the tour. I stopped walking long enough to close my eyes, turn my attention inward, and focus on my heart.

Following the guidance of Tibetan monks and nuns and yogis I've met, and certain of my indigenous friends, I knew that it would help to touch the area of my heart with my fingertips during this process to draw my awareness to the place of the touch. And as I began to slow my breathing, I felt a familiar sense of calm wash over my body. I felt like myself, and the more I did, the more the horrible events of the day began to take on a new meaning. As I felt the feelings of gratitude, in this instance, for the calm in my body and for the opportunity to make a powerful choice, I asked the question that no one else could answer. From a place of heart intelligence, without using my thinking mind, I silently asked, *Is this a good time for my group to experience the mysteries of Egypt?*

Through the years of using heart-based intelligence, I've learned that the heart works best when it's given brief phrases to respond to rather than multiple sentences. The heart doesn't need an introduction to the question that we're asking or an explanation of the history of the decision that's at hand. Our heart already knows all these things.

For some people, the wisdom of the heart comes as a feeling. For others, it can be a sense of knowing without question, while for still others, the answer emerges as a familiar voice that they've known throughout their lifetime. For me, it's generally a combination of all these. I often hear a subtle voice first, reinforced by a solid feeling of reassurance, safety, and certainty, which is followed by a sense of resolution and completeness. And that's precisely what happened on that day in the high desert.

Before I even finished asking the question, the answer was there for me, complete, direct, and clear. Immediately I felt—*I knew*—that our journey would be okay. It would be profound, deep, and healing. Most of all, I knew that by allowing our intuition to guide us through each step of our journey, we would be safe. I knew in that moment that I would soon be with my group in Egypt.

I want to be absolutely clear here about what I'm saying. *My decision to go forward with the trip was based upon the sensory impressions that I received as the result of a methodical and science-based process.* It was not made from a sense of hope that everything would go according to plan or by simply trusting that all would be well. While this kind of trust is perfect for some situations, when it comes to the lives and safety of a tour group the decision needs to be based on something more. For me, that something was the wisdom of deep intuition.

The steps that I applied to trigger my deep intuition also happen to parallel a process that other people sometimes use in a less structured way, but with similar results. The value of accessing heart intelligence is that it becomes possible to ask our questions with no attachment to the outcome, through *chante ishta*, the single eye of the heart.

When I was clear with my decision, I personally called each person scheduled for the tour to inform them. All of them, regardless of age or nationality, told me they trusted me to go forward with the trip, but only if I felt it was safe to do so—which I did.

Key 28: Intentional heart focus empowers us to consistently experience deep states of intuition when we choose, on demand.

THE REWARD FOR TRUSTING MY INTUITION

I left for Egypt on schedule the next week with 40 amazing people to begin a heartfelt adventure that would be full of surprises. We arrived in a country that was mourning the loss of many lives and reeling from the impact of the attack. The president of Egypt at the time, Hosni Mubarak, was a friend of our guide and extremely grateful that we came to his country during such a difficult time. We received an official letter from Mubarak giving the Department of Antiquities permission to open rare archaeological sites to us throughout our tour. Some of the sites, we later discovered, had not been opened to the public since they were first excavated in the late 1800s, and they have not been opened again since our tour! Needless to say, the journey was awe-inspiring, and the bonds between members of our group and the Egyptian people with whom they forged friendships have lasted to this very day.

The beauty of heart-based wisdom and choices made from that wisdom is that we're relieved of the burden of second-guessing our decisions. Based upon what I knew to be true at the time, I felt that my decision to lead the tour was a good one. I also believe that if I'd canceled the trip based upon what I knew that day, that also would have been a good decision. In having made the choice to go forward with the trip based upon the wisdom of my heart, however, I feel that I honored the people who trusted me to lead them, as well as myself, by making the best choice possible.

This story is only one example of how the tool of deep intuition has served me again and again in the real world. And while this example is about a big decision involving 40 people and a trip that took us halfway around the world, I use precisely the same technique, sometimes on a daily basis, to help me plan my schedule, temper relationships, and honor the principles that are important to me when I'm tested in life.

What I know with certainty is that we can never go wrong when we honor our hearts. I also know that if heart intelligence works for me, it will work for you.

YOUR HEART'S WISDOM IS TRUE ONLY FOR YOU

Your heart's intelligence is with you always. It's constant. You can trust it. It's important to acknowledge this because it means that the wisdom of your heart—the answers to the deepest and most mysterious questions of life that no one else can answer—already exist within you. Rather than being something that needs to be built or created before it can be used, the link between your heart and the place that holds your answers is already established. And while it's been with you since the time you were born, it's your choice as to when you access that link as a "hotline" to the deepest truths of your life.

You may choose to tap your heart's wisdom only in special circumstances, when there is nowhere to go and no one to turn to for guidance. Or you may choose to develop a relationship with your heart that becomes your second-nature, go-to source of guidance each and every day of your life. Regardless of the role you choose for heart intelligence in your life, it's up to you as to how you share what you hear from your heart and how you manage the reality of your everyday world. This is where discernment comes in.

While the guidance of your heart is true for you, it may not be true for another person. Our friends, children, siblings, life partners, and relatives all have their own heart wisdom to access. When we try to make a life-changing decision for other people in an isolated moment of time, we can't possibly know with certainty what's true for them in that moment. We can't possibly know, for example, intimate details of their life history from the time of their birth that have brought them to the present moment and their current circumstances. And because we cannot know these things with certainty, we can't anticipate how the well-intentioned sharing of our wisdom will affect the experience of another person.

I'm mentioning this now just as a point of consideration.

When you find yourself wondering if you should share what your heart has revealed to you, as a guideline I recommend asking the following three questions.

1. What is my intention in sharing what I've discovered?
2. Who will benefit if I share this information? Or more specifically: How will ________ benefit if I share this information? (Fill in the blank with the name of the person with whom you're considering sharing your revelation.)
3. Who may be hurt by my choice to share this information?

The key to using these questions is to be absolutely clear with yourself about the very first question. To be conscious of your intention is the foundation of your personal responsibility. With your intention firmly in place, it becomes easy to evaluate your answers to the next two questions to see if they honor your stated intention. Whether they do or don't, through this simple process you will find the answer to your question about the appropriateness of sharing your deep knowing.

With these ideas in mind, let's discuss how to apply the steps of coherence to access the intelligence and guidance of your heart.

ASKING YOUR HEART A QUESTION

Now that I've described the role of the heart in accessing deep intuition, I'd like to take this opportunity to share a proven technique that allows you to access its wisdom, as well. And I want this exercise to be personal, so I will offer this section as if I'm speaking to you directly while you are sitting with me in my living room. This exercise is one of those places where science and spirituality overlap beautifully. While science can describe the close relationship between the heart and the brain, the ancient spiritual practices and self-mastery techniques that have helped people rely on this relationship for thousands of years do so without needing a scientific explanation.

It's probably no coincidence that the rigorous scientific techniques developed by the researchers at the Institute of HeartMath closely parallel some of the techniques preserved in the monasteries of ancient traditions or by indigenous spiritual practitioners. We all learn in different ways, and my sense is that when something is true, it appears in the world in different forms to reflect the variations in our learning.

With this idea in mind, I've chosen to share the following IHM technique, with permission, because it's safe, it's based upon well-researched science that validates the steps, and it has been simplified in a way that makes it accessible and easy to use in our everyday lives.

As with any technique that's passed from teacher to student, however, the steps for creating heart-brain coherence are best experienced with a seasoned practitioner to facilitate the process. So while I'll describe these principles for creating heart-brain coherence in the following paragraphs, I also encourage you to experience them for yourself using the no-cost online instructions found on the Institute of HeartMath website (see the Resources section).

The technique to create heart-brain coherence is appropriately called the Quick Coherence® Technique and has been refined by the Institute of HeartMath into the first three simple steps described below. Independently, each step sends a signal to the body that a specific shift has been put into motion. Combined, the steps create an experience that takes us back to a natural harmony that existed in our bodies earlier in life, before we began to disconnect our heart-brain network through our conditioning. Steps 4 and 5, where we access our heart's wisdom, build upon the coherence created in Steps 1 through 3.

Five Steps to Ask Your Heart a Question

The steps to create quick coherence for accessing your heart's intelligence are as follows.

- **Step 1: Create Heart Focus**
  - **Action:** Allow your awareness to move from your mind to the area of your heart.
  - **Result:** This sends a signal to your heart that a shift has taken place: You are no longer engaged in the world around you and are now becoming aware of the world within you.
- **Step 2: Slow Your Breathing**
  - **Action:** Begin to breathe a little more slowly than usual. Take approximately five to six seconds to inhale, and use the same pace as you exhale.

  - o **Result:** This simple step sends a second signal to your body that you are safe and in a place that supports your process. Deep, slow breathing has long been known to stimulate the relaxation response of the nervous system (aka the *parasympathetic response*).

- **Step 3: Feel a Rejuvenating Feeling**
  - o **Action:** To the best of your ability, feel a genuine sense of care, appreciation, gratitude, or compassion for anything or anyone. The key to success here is that your feeling be as sincere and heartfelt as possible.
  - o **Result:** The quality of this feeling fine-tunes and optimizes the coherence between your heart and your brain. While everyone is capable of evoking a feeling for this step, it's one of those processes that you may need to experiment with to find what works best for you.

With the successful completion of Step 3, the connection linking the heart and brain—and resulting in heart-brain coherence—has been established. At this point, the heart and brain are in communication through the neural network that connects them. While this is technically the completion of the Quick Coherence® Technique itself, it's also a beginning step in other processes. We may use the coherence we've created to access deeper states of awareness, including the deep intuition described in this chapter. It's from a state of heart-brain coherence that we may access our deep intuition and receive the guidance of our heart's intelligence. Steps 4 and 5 below detail a procedure to do just that.

- **Step 4: Ask Your Heart a Question**
  - o **Action:** The previous three steps create the harmony between your brain and your heart that enables you to tap into your heart's intelligence. As you continue to breathe and hold the focus in your heart, it is time to ask your question.

    Heart intelligence generally works best when the questions are brief and to the point. Remember, your heart doesn't need a preface or the history of a situation before the question. Ask your question silently, as a single concise sentence, and then allow your heart to respond in a way that works for you.

- **Result:** Your intuition opens up and you begin a dialogue.

I'm often asked to interpret the symbols that show up in people's dreams or the meaning of an experience that they've had in their lives. While it's possible for me to offer an opinion, it's just that. It's *my* sense of what the image or experience may mean in *their* life. The truth is that I can't possibly know what another person's dream or experience means for them. It's also true that they can!

The key to being successful at dialoguing with your heart is this: *If you are empowered enough to have the experience, then you are empowered to know for yourself what your experience means.*

While I don't want to influence your questioning process, an example is sometimes helpful. A mysterious dream is the perfect opportunity to apply heart wisdom to a real-world situation. From the heart-brain coherence established in the previous three steps, simply ask the following kind of questions, filling in the blank with the names of the people, symbols, or identities of what you're asking about. These are example formats only. You can choose one that fits for you or create your own using one of the following as a template.

- "From the place of my heart's deepest knowing, I ask to be shown the significance of ________________ in my dream."
- "From the single eye of my heart that knows only my truth, I ask for the meaning of the ________________ that I saw in my dream."
- "Please help me to understand the significance of ________________ in my life."

- **Step 5: Listen for an Answer**
  - **Action:** Become aware of how your body feels immediately as you are asking your question in Step 4. Make a note of any sensations—such as warmth, tingling, or ringing of the ears—and emotions that may arise. For people who are already attuned to their bodies and their hearts' intelligence, this step is the easiest part of the process. For those who may have had less experience in listening to their bodies, this is an exercise in awareness.

- o **Result:** Everyone learns and experiences uniquely. There is no correct or incorrect way of receiving your heart's wisdom. The key here is to know what works best for you.

As I mentioned before, I tend to receive my heart's wisdom as words, while at the same time feeling sensations of warmth in my body. Other people never hear words but experience nonverbal forms of communication only, such as warmth radiating from their hearts or in their guts. Sometimes people feel a wave of peace wash over them as they receive the answer to their question. Remember, you and your body are unique partners in the world. What's important here is to listen to your own body to learn how it communicates with you and give it the opportunity to be heard.

Now you have a step-by-step technique to help you feel empowered in the face of life's greatest challenges. While you probably can't change the situations that arrive at your doorstep, you can definitely change the way you feel in and respond to those situations. If you've not already done so, you may discover that the wisdom of your heart becomes a great friend to you, one of the greatest sources of strength in your life. The consistency and accuracy of heart-based solutions empowers you to face any situation with any person or force with a confidence that's hard to find if you feel helpless, overwhelmed, powerless, and lost.

I can honestly say that my heart's wisdom has never led me to make a bad choice. And while I haven't used this technique for every big decision I've made in my life, I can also say with honesty that the only choices I've regretted are the ones I made when I did not honor my heart's wisdom.

As you complete this exercise, I invite you to bear an important point in mind: There is no correct or incorrect way of receiving your heart's wisdom. Each of us is born with our own unique code that allows us to access our heart's wisdom and apply it in our lives. The secret to the code is to know what works best for you.

Key 29: We can access our heart's wisdom through a process that can be summarized in five simple steps: focus, breathe, feel, ask, and listen.

IT'S SECOND NATURE TO ASK YOUR HEART

Your intuition can help you feel more empowered in the face of life's greatest challenges. Each time you access your heart's wisdom, you are actually reinforcing and strengthening the neural connections that make our heart-brain connection possible. I commonly hear from people who incorporate heart intelligence into their daily lives that the Quick Coherence® Technique becomes easier to do over time.

In fact, for some people, the experience becomes second nature, so that for them it is an automatic response rather than a structured technique. They find themselves instinctively shifting their awareness to their hearts multiple times throughout the day to gain perspective on life's challenges and to balance life's demands. They also discover that once they are in the heart, the ability to embrace life issues in a compassionate way becomes second nature as well.

When people share their stories of such experiences, although I'm always in awe of the process, I'm not surprised by what I hear, because the intuition that naturally flows from our hearts provides us with a stepping-stone to experience the deepest levels of sympathy, empathy, and ultimately compassion in our lives. When we think about it, this flow of experience makes perfect sense. After all, how can we relate to someone in a compassionate way if we cannot first identify with the suffering they experience—and do so in a healthy way? The ability to identify with another person's experience of hurt, distress, or trauma without taking on their suffering as our own—an experience that is sometimes called *overcare*—is the key to effectively supporting someone in their pain, distress, and trauma. This is where empathy comes in.

The ability to relate to another person—or any form of life, for that matter—on an intimate level is known as *empathy.* Our capacity for empathy is the key to our capacity for compassion.

EMPATHY: A STEPPING-STONE TO COMPASSION

In the popular TV series *Star Trek: The Next Generation* (1987–1994), one of the core characters is counselor Deanna Troi (played by Marina Sirtis), an *empath*—a person with the ability to sense the feelings and emotions of another being while experiencing them on a personal level. Knowing that the stated mission of their futuristic journey through the universe is "to explore strange new worlds, to seek out new life and new civilizations, and to boldly go where no one has gone before," it makes perfect sense to have a skilled empath as an integral part of the ship's crew. The multiyear duration of the *Enterprise*'s mission makes it likely that the crew will encounter forms of life that do not communicate using words as we humans do. And throughout the series, that's precisely what happens. Thanks to the empathic skills of counselor Troi, however, those nonverbal exchanges are not a problem. Although each encounter with an alien species is unique, such encounters tend to follow a common theme that goes something like the following plotline.

The captain of the *Enterprise* is communicating with the leader of an alien craft that has suddenly appeared with unknown intentions. While the leader of the dark ship says to the captain, "We come in peace," with his words, counselor Troi is sensing another intention nonverbally. As an empath, she feels that something dangerous lies beneath the exchange between the two leaders. So while the *Enterprise*'s captain is listening to the alien, counselor Troi is whispering what she senses into the captain's ear, such as:

"They want to destroy us." In this instance, it's easy to see why the counselor role is so valuable for the mission of the *Enterprise.*

Although the TV series is science fiction, the empathic abilities of counselor Troi are not. They're real, and we each experience them to a greater or lesser degree in our daily lives, often without even realizing what we're experiencing.

So what is empathy? How is it related to sympathy? And how may we experience both in a healthy way?

Both empathy and sympathy are forms of intuition, and the words that describe these states share a common origin in the Greek language. They come from the root word *pathos,* which means "feeling." This is where knowing a little Greek offers us a clear distinction between their meanings. The Greek prefix *sym-* in *sympathy* means "with." The prefix *em-* in *empathy* means "within." From translating these simple prefixes, the difference becomes clear.

To have *sympathy* means to identify *with* others in their hurt or suffering. When we're sympathetic, we say that we feel for the plight or the loss of another person. When friends or family members experience the death of a loved one, for example, we send sympathy cards to let them know that we are acknowledging their loss and sense what it must mean to them.

When we're sympathetic toward other beings, we're observers, longing to stand with them and support them in their experience. We sometimes say that we can "only imagine" what such a loss may feel like. And when we say this, our statement is completely accurate. Because the loss we're sympathizing with is not ours directly, we are left to identify with our loved ones' hurt by drawing upon memories of our own experiences to approximate what they must be feeling. Sympathy is the first step to achieving empathy.

When we experience *empathy* for others, we go beyond sympathy. We begin to close the gap between acknowledging someone else's suffering from a distance and feeling their suffering ourselves. We place ourselves perceptually and emotionally in their situation to experience what they are experiencing. In doing so we identify more closely, in an even more profound way, with the suffering of others.

Both sympathy and empathy are precursors of *compassion.* We must first experience empathy for another person's suffering before we can become compassionate in the way we respond to them. Just to be clear, however, having empathy in a situation doesn't necessarily mean that we will become compassionate. It's possible to have empathy for another person's experience without that empathy leading to compassion.

To be a compassionate person is a choice. And when we make such a choice, our experience carries us to a deeper level of experience.

Key 30: Intuition, sympathy, and empathy are the steppingstones to compassion.

In compassion, we become involved. *We actually do something in an attempt to alleviate the suffering of one or more others.* And while we hope our actions will ultimately contribute in some way to alleviating the suffering of others, the goal of compassion is less about the outcome itself and more about us and how we are changed in the presence of a compassionate choice. Once our lives reflect the compassion that we choose, it's then that the compassion we've become can be reflected in everything we do.

For centuries, the great spiritual masters have reminded us that a compassionate response to our world begins with us and lives in the way we relate to the world. From this perspective we can say that compassion is a powerful inner technology, an advanced form of intuition that gives us the power to create sustainable solutions from a very personal level.

Any doubt I originally had regarding the power of compassion in our lives disappeared after I had the opportunity to spend time with Tibetans who had been steeped in the traditions of compassion from an early age.

MEETING WITH THE ABBOT

On an icy high-altitude morning in the spring of 1998, I found myself living a reality that I'd dreamed about for as long as I could remember. I was leading a combined research trip and pilgrimage into one of the most magnificent, pristine, remote, and absolutely beautiful places remaining on earth, the rugged Tibetan Plateau, a place where Buddhist monasteries have survived the harsh elements of time for over 1,500 years.

On the 16th day of the trip, I found myself seated along with a few members of my group in the cramped quarters of a tiny chapel hidden deep within the massive walls of the ancient monastery we were visiting that day. Surrounded by Buddhist altars and faded *thangkas* (intricately brocaded tapestries that preserve the great teachings of the past), which were barely visible in the dim light, we sat face-to-face with the highest-ranking religious leader of the monastery, the elder abbot. Through the skills of our translator we were given a private audience with this lifelong student of meditation and compassion.

During the hour or so of this intimate encounter, I had the opportunity to ask questions about Tibetan traditions, beliefs, and the deepest mysteries of life. My questions were direct and to the point, and the abbot clearly enjoyed the opportunity to vary his daily routine with our meeting. So much so, in fact, that he even resisted the urging from his assistants who were trying to help him leave us to go to another appointment.

I'm sharing this story because it was the trust and friendship from this initial meeting that paved the way for a second meeting in a different chapel of the same monastery seven years later.

In 2005, I had the opportunity to revisit the monasteries of the Tibetan Plateau. This time I was with another group of researchers and pilgrims, and our trip lasted a total of 18 days. When we returned to the monastery that I'd visited previously, we learned that the elder abbot who had been so generous with his time in 1998 was no longer there—he had died. Although we never learned the details of when or how he'd passed, the monks left no doubt in our minds that he was no longer in this world. Clearly, however, the friendships that we'd established seven years earlier had left a mark of goodwill among those who had been assistants to the elder abbot and among other monks still living in the monastery. So even though we'd never met the new, younger leader—only in his late 80s—who had replaced the elder, our goodwill and strong relationships preceded us. When the new abbot heard that our group had returned, we were welcomed warmly and allowed another opportunity to continue the deep conversation that began seven years before.

THE FORCE THAT CONNECTS ALL THINGS

On another frosty Tibetan morning, this time in a different chapel of the same monastery, we found ourselves seated with the new abbot. Only minutes before, we'd been led through a meandering, stone-lined passageway that led to this tiny, cold, and dimly lit room. As we waited for the abbot to meet us, I remember thinking to myself that we could only imagine the conversations, teachings, and initiation processes that had occurred in the very place we found ourselves that morning. In the distance, I heard the faint sound of leather sandals slapping against the cold stone floors. I knew that it was the abbot coming for our meeting. As the sound grew louder, I could feel the growing anticipation in the room with the realization that our meeting, though delayed, was really going to happen.

The abbot pushed aside the heavy tapestry that hung in the doorway to keep the cold air out (or to keep the warmer air in the room). With a huge grin, he touched the thumb of his right hand to his heart, fingers together and pointing toward the sky, in a half prayer mudra, while the other hand held on to his robes as he glided across the room. Following the formalities of introductions and the blessing of the *khatas*—ceremonial white silk scarves that each person who meets him traditionally offers him for blessing—the abbot signaled that he was available for questions. It was here, nestled in the silence of the ancient monastery, that I asked a question regarding the topic of the book I was writing at the time, *The Divine Matrix*.

"In your tradition," I began, "what is the force that connects us with other people, our world, and all things? What is the conduit that carries our prayers beyond our bodies and the stuff that holds the universe together?"

With the smile that never left his face, the abbot held eye contact with me as our translator repeated my question in Tibetan. What happened next came as a surprise to me and the others in the room.

Immediately the two men—the abbot and his translator—began a loud and lively exchange with animated gestures and enthusiastic emphasis that began to sound like a Tibetan shouting match! While my Tibetan is terrible and I couldn't understand a single word that either man was saying, the nature of the conversation seemed obvious. They were struggling with the meaning of my question and where it fit into the abbot's teachings. He was accustomed to answering such questions from familiar students who had already studied with him and had years of training that would prepare them for such a conversation. But the abbot didn't know me. He had no idea of my background, my traditions, or my spiritual experience, and he simply didn't know where and how to begin his answer.

Answering me as he might answer a lifelong monk would be like parents telling a young child where babies come from without the child first having the benefit of knowing about the biology of intimate human relations. While such a question could be answered, for the child the answer wouldn't make sense without the prior knowledge. In a similar way, the abbot knew he could answer my question regarding the force that connects all things. What he didn't know is if I would understand his answer.

A FORCE OF NATURE, A HUMAN EXPERIENCE, OR BOTH?

Suddenly the room became quiet. Everyone stopped talking and the abbot lifted his gaze upward to the thangkas that covered the chapel walls. After inhaling deeply the cool, thin air, he answered my question in a way that was both surprising and unexpected. He looked at me and simply spoke one Tibetan word. Instinctively I glanced to the translator. "What did he just say?" I asked. "It was only one word!"

I wasn't prepared for what I heard next from my translator. "Compassion," he said. "The *geshe* says that 'compassion' is the answer to your question. Compassion is what connects us to every creature and to all things."

The reason I was surprised by the answer is because in my experience, compassion had been taught as an experience and a practice. We *feel* compassion for ourselves and for others who are facing difficult life circumstances. We also *experience* compassion as a practice in our daily lives. If I understood the abbot's answer correctly, he was now telling us that compassion is more than a feeling—it's a force of nature.

I had never heard compassion referred to as a physical force. Yet this single word was his answer to my question: "What connects us with our world?" And this apparent contradiction was the source of my next question. "How can that be?" I asked the translator, looking for clarity in what I was hearing. "Is he telling us that compassion is a *force of nature* that connects all things, or is he telling us that it's a human *emotion that we experience?*"

Once again an animated dialogue broke out as the translator conveyed my question to the abbot. Once again, the abbot broke his connection with my eyes, took a deep breath, looked toward the translator, and answered my question with a single word. "Yes!" he said in Tibetan. That was his answer. It was also the end of our exchange.

Following nearly 10 minutes of a bantering dialogue involving the deepest elements of Tibetan Buddhism, all that I got to take away with me was the Tibetan word for *compassion.* I remember leaving the monastery that day feeling incomplete, like there was something that had been literally lost in translation. The answer from the abbot was a bit mysterious and didn't seem to make sense. Something just didn't fit.

A few days later I discovered the reason why.

At yet another monastery, this time with a scholarly monk rather than a high-ranking abbot, I found myself engaged in the same conversation once again. This time, however, we were in the casual environment of the monk's cell. It was the tiny, unadorned room where he ate, slept, prayed, and studied when he was not in the monastery's great chanting hall.

By now, our translator was becoming familiar with the form of my questions and what I was trying to understand. As we huddled around the heat of the yak-butter lamps burning in the smoke-filled room, I looked up at the low ceiling. It was covered with black soot from countless years of similar lamps burning for heat and light, just the way they were on that cold afternoon.

Once again, through the translator, I posed the same question to the monk: "Is compassion a force of creation or is it an experience in the body?" He rolled his eyes to the soot on the ceiling where I had been looking only seconds before. With a deep sigh, he thought for a moment, drawing from what he had learned in the monastery since he entered at the age of eight. He now appeared to be in his mid-20s. Slowly he lowered his eyes and looked at me as he responded.

The answer was short. It was powerful. And it made tremendous sense. "It is both!" were the words that came back to me from the monk. "Compassion is *both* a force of nature *and* a human experience."

In that moment the earlier encounter with the abbot suddenly made sense, and I understood the deep teaching that he had shared with me and the members of my tour group.

Key 31: Compassion is both a force of nature and an emotional experience that connects us with nature and all life.

Einstein's Compassion

On that day in a monk's room halfway around the world, hours from the nearest town, at nearly 15,000 feet above sea level, I heard the words of a simple, yet powerful, wisdom that many Western traditions, including science, have overlooked to this day. The monk had just reminded us that a single human experience that sets us apart from all other forms of life—compassion—is the same force of nature that intimately connects us with all things. When we experience true compassion in our lives, our sense of separation between ourselves and others, all life, and the world, as well as within ourselves, disappears.

Albert Einstein recognized the power of compassion in our lives, as well as the potential it holds to alleviate suffering. In his words, "Our task must be to free ourselves . . . by widening our circle of compassion to embrace all living creatures and the whole of nature in its beauty."[10] The fourteenth Dalai Lama carried this understanding from personal healing to global survival, stating,

"I truly believe that compassion provides the basis of human survival."[11] The recognition of compassion's role in our lives opens the door to the depths of our greatest self-mastery and the extraordinary experiences that make us human.

As a masterful teacher, the abbot felt a responsibility to answer the questions from his students in a way that is both honoring and meaningful. Without knowing anything about me, my background, or my history and beliefs, the abbot had no way of knowing if his wisdom would be either honoring or meaningful for me. He simply didn't know how his words would land in my life experience. That was the root of the struggle that I witnessed between him and the translator before he spoke the word *compassion.*

Fortunately for me, my translator was also a good friend. He knew me. He knew about my family, my life, my corporate and academic background, my education, and my spiritual journey. Armed with that knowledge, he was able to reassure the abbot that whatever wisdom he chose to share would find its way to my mind and to my heart in a respectful and healthy way. That was everything the abbot needed to hear to reassure him that he was honoring his sense of responsibility. In doing so he had expanded the way I'd been taught to think about compassion and the role it plays in our lives.

COMPASSION, WISDOM, AND BALANCE

The teachings of our Tibetan abbot specifically, and of Tibetan Buddhism in general, are based in the Mahayana Buddhist traditions, one of the two (or in some classifications, three) major branches of Buddhism. According to its teachings, Mahayana is the path that quickly leads an individual to complete enlightenment for a single purpose: so that they may use their enlightenment to alleviate the suffering of others. Someone who follows such a path is known as a *bodhisattva*. I'm sharing this background here because of the context it offers for understanding compassion.

The sensual nature and poetic language of the Mahayana teachings (known as *sutras*) have always touched me with their beauty. They have also been a refuge and a source of comfort in some of the most difficult times of my life. When it comes to the sutras' description of compassion, for example, they portray bodhisattvas as having two wings that carry them toward the goal of enlightenment. One wing is the wing of wisdom. The other wing is the wing of compassion. The sutras describe these two qualities as equals in a partnership that is necessary for each of us if we choose a journey to enlightenment.

In an especially powerful way, the sutras describe how the bodhisattvas have no place to stand in the world. The reason is that there is nothing that bodhisattvas can call their own. They have no land, no possessions, and no attachments in the world. But this idea goes even deeper, penetrating the essence of the way we think of ourselves in the world. This deeper layer of the bodhisattva is best described in the words of Buddhist scholar Joanna Macy, Ph.D. "Nor is there a solid self, or an unchanging identity, or any security, as we understand security," she says.[12] Instead, bodhisattvas move confidently through the world, trusting in the wisdom and compassion they have attained to navigate any situation that life may put at their feet.

The key insight here is that compassion must be balanced with wisdom to serve us in a healthy way. Compassion and wisdom must become our allies if we are to express the deepest truths of our humanness.

My journey to unlock the mystery of intuition and compassion in my own life has led me to some of the most secluded and mysterious places remaining on earth. It's in the ancient monasteries and nunneries, within the brittle pages of timeworn manuscripts, and among the indigenous people themselves that their wisdom has been preserved for us today. Rather than discovering cut-and-dried answers in these places, what I've found are the keys to a way of thinking that makes new answers and new ways of thinking possible. Maybe it should come as no surprise that the clues to our bodies' deepest mysteries and to our greatest powers are hidden in the plain sight of our everyday experiences. And the mystery doesn't end when we learn the language of the heart.

Just like the introduction of a book is the guide that prepares us for the chapters that follow, our sense of intuition and compassion guide us through the nuances of life and give us a means to answer the questions that show up in everyday living.

chapter six

We're "Wired" for Long Life

Awakening the Power of Our Immortal Cells

"Many things can prolong your life, but only wisdom can save it."

— Neel Burton (1978–), British psychiatrist and philosopher

"From the moment we're born, we begin to die."

I was hearing these words from a dear friend whom I'd known in northern Missouri during my teen years. (I'll call him Michael to honor his privacy.) We each shared histories so similar, it sounded like we could be brothers. Both of our fathers had left our families when we were 10 years old. We both had a younger sibling. We both found ourselves living in the same low-income government-subsidized housing and walking to the same school and the same classes each morning. And we'd both turned to music to help us cope with the turbulent world of broken homes that we found ourselves in, with me as a guitarist and Michael as a drummer. The year was 1968 and together we'd witnessed on live television the aftermath of the assassinations of Martin Luther King, Jr., and then Robert Kennedy barely two months later, along with the horrors of the killing of protestors at the South Carolina State University campus and police brutality during anti-Vietnam riots during the Democratic National Convention in Chicago. We played together in the same rock group, and after our already late-night band practices, we'd stay awake until the wee hours of the next morning talking about America, American politics, and the future of the world.

It was against the backdrop of this friendship that Michael shared his life philosophy that *we begin to die the moment we're born*. While I was certainly familiar with the saying, when I'd heard it in the past I'd typically written it off as a fringe idea—one that I didn't necessarily agree with—but accepted as one of the many new viewpoints that were emerging at the time. When I heard it from my friend, however, something felt different. This time the words were coming from someone I cared about, and they were being used to justify a way of thinking and living that paved the way to a life of excess, overindulgence, and ultimately, bad endings.

Michael and I were deeply engaged in a conversation about life and how to live it to the fullest. It was a conversation that couldn't have come from more different perspectives. Michael believed what he'd heard about us dying from the minute we're born, and he had taken this to heart as his primary life philosophy. He literally believed that our lives are like a sealed jar of potential. When we're born, we open the jar and begin to use up our potential from the instant of our first breath.

"We have what we have, and when it's over it's over," he said. "When it's gone, it's gone."

ARE WE LIVING IN THE MOMENT OR ARE WE LIVING FOR THE MOMENT?

In Michael's way of thinking, there are two mysterious unknowns that are with us when we begin our lives. The first is that we simply don't know how full our "jar of life" is when we arrive in this world. The second is that we don't know how fast we'll use whatever life we're given. We could be blessed with a jar that's overflowing with vibrant health. If so, we could be in this world for 100 years or more. Or we might begin our lives with our jar only half full—something that Michael called "beginning with half a tank." His thinking was that if we started with less, we'd use what we had faster, life would be short, and we'd die young. Michael's belief was that it's precisely *because we don't know* how full or empty our own life vessel is that it makes sense to live life full-on, for the only moment that is certain: the moment of now.

While I got his meaning and understood the underlying philosophy of what my friend was saying, I also know that this kind of thinking means different things to different people. For Michael, the idea of living *for* the moment meant speaking whatever words came to mind and acting in whatever way moved him in any given moment. (This is very different from living *in* the moment, where we fully embrace our senses, we become aware of our surroundings, and we live, act, and speak consciously and responsibly from our enhanced awareness.) From Michael's point of view, he felt that in order to honestly live a spontaneous life, there could be no filters on what he said or what he did. Each moment simply *was* what it was. And it was precisely this thinking that had prompted our conversation.

Not surprisingly, Michael was in full-blown life-crisis mode when we met that day. His interpretation of living *for* the moment had led him to avoid all commitments at all costs: commitment to himself, to his family, to his body and his health, to other people, to friendships, and to intimacy. The consequences of his approach to life had caught up with him and brought him disappointment, unfulfilled dreams, and the sense that, while a successful, healthy, and loving lifetime was certainly possible for some people, that possibility was for everyone except him.

In the moment of our conversation, Michael found himself experiencing a health crisis. He was only in his twenties, yet the intensity of his drug and alcohol use had culminated in a liver condition that needed immediate medical attention. He was also alone in his life. He had no money, nowhere to live, and no one to turn to. As far as I knew, I was his one remaining friend. And while I've learned to hold off when it comes to offering personal advice to friends (unless I'm asked to share), my friend seem to be mired so deeply in his pain that it never occurred to him that there could be another way of thinking about his life philosophy, and it didn't occur to me to withhold. I took a chance and offered some counsel. "What if the life philosophy you heard of, about dying from the moment we're born, is a bit off?" I asked.

The look on Michael's face when he heard my question told me I had his attention. "What do you mean when you say 'a bit off'?" he said.

"I was being kind," I replied with a smile. "I didn't want to shoot down your entire worldview in a single sentence."

"Okay, I got it!" he said. "What are you telling me? Just say the words."

Hearing my friend ask me for clarity was just what I'd been hoping for. It was the way in, my opportunity to offer another point of view, and I grabbed that opportunity immediately. "What if life works in a way that is just the opposite of what you've been led to believe?" I asked. *"What would it mean if you discovered that from the moment we're born we begin to heal?"*

Michael looked stunned. This simple shift of thinking had never occurred to him. Just hearing the words opened the door to a possibility that he'd never even considered. "Wow! If that were true," he said, "it would change everything. It would mean that we can keep filling our tank of life forever—or at least for a long time."

"I know," I said. "That's the whole point. And we don't need to ask what it would mean *if* we discovered this possibility, because we already have. Ancient traditions, such as yoga, qigong, and Ayurvedic medicine, have already discovered that our bodies are literally wired to heal from the instant we're born. They've also discovered that we're the ones that start and stop the process. The key is to create the conditions that make the healing possible. There are many ways to do so, and this is why we can think of life in terms of a tank that we keep filling, rather than one that becomes emptier with each passing day."

Shortly after our conversation, Michael moved to another city. His estranged father heard about what was happening, contacted Michael, and offered him a home while he was working out his health issues. Over the years I lost track of my friend. I never saw him again. But when I think back to those years in northern Missouri, I'm always grateful for the deep conversations that gave us both new ways to see the world, and to think about our lives.

The more I've learned about the wisdom of the human body, and the more time I've spent with indigenous people that embrace that wisdom, the more I've come to understand about the potential that I shared with Michael. The capacity to heal is something that we already have and that already lives within each of us. From yogis, monks, and nuns to shamans, mystics, and *curanderos*, people whose respective traditions are very different from one another, there's a fundamental theme that weaves the threads of each tradition into a single, powerful tapestry. These ancient and indigenous traditions teach that the quality of our lives, and how long we live, at root comes down to the way we think of ourselves in the world.

To replace the limiting beliefs that we've learned from our families, friends, and social institutions with new, self-empowering perspectives truly does, in Michael's words, "change everything." I discovered this for myself in the most direct way possible when

I met a Tibetan nun who defied the conventional wisdom I had learned as to what age means and the role of longevity in our lives.

THE NUN'S SECRET

After nearly two weeks of acclimating to elevations as high as 16,000 feet above sea level and having our bodies bounced on the stiff spring seats of an ancient Chinese school bus along roads that were little more than washed-out trails, we arrived at the remote monastery. It was hours away from the nearest village and occupied only by a group of about 100 Tibetan nuns who had little contact with the outside world and welcomed very few visitors. The dust billowed into the air above the surrounding hills, alerting the nuns in advance that we were on our way. They were waiting for us as we arrived, standing quietly among a crowd of curious but shy children, local farmers and yak herders, and weather-hardened nomads.

It seems that every photo opportunity in Tibet was what our group members called a "*National Geographic* moment," meaning that a picture capturing it would qualify as a cover image for the popular magazine. This moment was no different. Three of the nuns immediately stepped forward and, following a few warm greetings, informed us that they would be our official guides. These women were dressed in traditional nun's clothing: a deep maroon wrap (*zhen*) covering a maroon skirt for the lower body (*shemdop*) and a yellow and maroon wrapping shirt (*dhonka*) covering the upper body. The big smiles on their faces and their animated conversation told me that they were excited for the opportunity to meet with us.

Through our translator, the nuns confided that living in such an isolated location had its pros and cons. On the one hand, the monastery compound was so remote that government inspectors and land speculators rarely made the effort to disturb their religious community. On the other hand, the nuns were so far from the nearest town, so isolated, and the road to get to the monastery was so bad, that the tourism that would normally support the economy of such a place was almost nonexistent.

Our bus could seat exactly 40 people, plus one guide, a translator, and me. Needless to say, the sight of 43 people converging onto the monastery was welcomed and immediately brought the courtyard to life as sales kiosks magically sprang up for business everywhere the eye could see. For an hour or so, our group did our best to support the local economy. We were true consumers, buying beautiful Tibetan rugs, brilliantly colored thangkas, and religious aids ranging from long ropes hung with prayer flags and huge Tibetan "singing bowls" made of brass that could be played with a mallet to strands of prayer beads used to count the number of times a chant would be repeated.

Suddenly the entire scene changed. As if they were following some sort of internal cue, the yak wool rugs, turquoise jewelry, singing bowls, and paintings were packed into big woolen bags, the kiosks were dismantled, and the nuns began walking in silence toward the buildings. "We're going now to the chanting hall," our tour guide whispered to me as I looked at him for an explanation. "It's time for the nuns to pray."

As we walked along a narrow path that was carved into the side of the mountain, one of the nuns came to walk beside me. I was immediately fascinated by her presence. She was close enough to me that I could easily estimate her height. My mom is exactly four feet eight inches tall, and the nun's face was level with the same place on my chest as my mom's when we took our walks together. But it was more than her height that fascinated me.

ONE HUNDRED TWENTY YEARS ON EARTH, BUT WHO'S COUNTING?

This woman's eyes were clear and bright, and she had a smile on her face the entire time we were walking together. Though the skin on her face looked healthy, I knew that the deep wrinkles around her eyes and across her forehead could be present only after a full lifetime of exposure to high-altitude sun, nature's elements, and the challenges of being a woman in such a harsh environment. Her head was freshly shaven, but I learned this was due to the lack of indoor plumbing, which made full-body bathing impractical, rather than to loss of hair due to her age. As we walked, it became clear that my Tibetan language skills were far worse than her limited English, and we quickly realized we wouldn't have much of a conversation. Not a verbal one anyway. Together we walked in silence to the prayer room as she alternated her gaze between looking down to the path and then up to meet my eyes.

When we reached the door to the chanting hall, with a quick nod the nun pulled back the heavily brocaded tapestry that kept the wind, dust, and bright sunlight out of the prayer space. My new friend stepped inside first, and before I could follow, our guide stopped me for a moment.

"Did you enjoy your conversation with the *geshe*?" he asked. *Geshe* is the Tibetan word for a great teacher, and while I sensed that the woman who had walked with me was a respected elder, I didn't know why she was held in such high esteem. Furthermore, while my guide casually referred to this nun as *geshe*, in Tibetan Buddhism this title has traditionally been reserved for highly educated *males* only. It was only in 2011, three years after our trip to Tibet, that Kelsang Wangmo made history by becoming the first officially recognized female geshe, and in doing so signaled a new era of possibility for women in Tibetan Buddhism.

I wasn't prepared for what I heard next. "The nun you were walking with holds the memory of this place, and the tradition of these women," my guide explained. "I called her the geshe because not only does she *know* the history, she actually *remembers* the history."

"What do you mean 'she remembers the history'?" I asked. "How is that possible? How can she remember what's happened in this place for over a hundred years?"

"That's why she is the geshe," he replied with a grin. Then he looked directly into my eyes and revealed the secret of why he wanted me to meet the woman I had just walked the trail with.

"The nun that you just walked with," he said, "remembers the history because she lived the history. She was born here in 1888 and has lived in this village all of her life." At first I thought my guide was teasing me. It quickly became clear that he wasn't.

"Yes," he said. "The mother superior showed me her papers. The nun is one hundred and twenty years old this year." (The year of this particular trip was 2008.) "And she's not the oldest of these people," he continued. "There are others here in the mountains that are much older."

"How much older?" I asked.

"That's the problem," he said. "The oldest ones are men who are now yogis. They live in the caves between Lhasa and the sacred mountain, Mount Kailash. According to the local villagers, some of them are six hundred years old! The problem is that they did not have birth certificates and passports six hundred years ago. We cannot prove their age with certainty."

And that's precisely why I valued my meeting with the Tibetan nun, and those that knew her, so much. Her exact age was known, and it could be documented because her records had been preserved in the monastery's library. She was still very much alive, very vibrant, and very happy to talk about her long life and the secret of achieving it. It was through the translator that afternoon that I asked what she felt was the secret to her longevity.

My new friend didn't miss a breath as she answered. Quickly, as if on cue, her answer was simple, brief, and concise. It left no doubt in my mind as to what she was telling me. "Compassion," she replied. "Compassion is life. It's what we practice here. It's what we learn from our masters and what they learn from theirs. It's what's written in these books." She gestured to the ancient and tattered manuscripts stored in the monastery library as she spoke. "It's what we keep safe to share with those who come here to learn."

With the recent discovery of a biological clock inside each cell that sets the "timer" for how long we live, her answer now makes perfect sense.

RETHINKING THE LONGEVITY PARADIGM

It may be no coincidence that the oldest documented ages for people in the world today seem to hover around the age of the nun I met in Tibet. They are at, or near, the 120-year mark. While there are certainly exceptions with some people that are just under this age, and some that are over, 120 years seems to represent some kind of a mysterious boundary when it comes to human longevity. From a biblical perspective, it hasn't always been this way. If we're to believe the accounts of the Hebrew Torah (subsequently the Christian Old Testament), for example, the biblical patriarchs lived life spans measuring multiple centuries rather than mere decades.

Methuselah, for example, was 187 years old when he fathered his son Lamech. So he was obviously a vital 187 years old at the time he did the fathering. Contradicting the way we've been led to think about longevity and how vitality diminishes with age, Methuselah lived another 782 years, and during that time he fathered additional sons and daughters, for a stunning life span of 969 years! And Methuselah is not alone in his longevity. The same biblical traditions tell us that at the age of 500 years, Noah "begat Shem, Ham, and Japheth."[1] So once again, to father three children, we know that Noah had to be both vital and virile.

These two accounts portray very different ideas of longevity from what we're conditioned to expect today. Our society and culture have programmed us to expect an inverse relationship between age and human potential. That expectation goes like this: The longer our lives continue, the fewer capabilities of our youth are available to us. A corollary to this expectation is the idea that while we're alive the quality of the life that's available to us diminishes with the passing years.

It's for these reasons that when we think of someone being 100 years of age or older, we are conditioned to think of them as a shell of their former selves. The image is one of a shriveled little human with muscles that have lost their tone sagging from the brittle bones of a body with dull and empty eyes, clinging to the last possible breath of life. And while this way of aging is certainly possible, and we've all seen this possibility realized somewhere among our families, friends, and neighbors—and clearly there's nothing wrong with accepting this possibility—the point I'm making here is that there's another possibility. It's a real possibility of vital longevity and it's more than a wish or a pipe dream. We also have seen examples, both ancient and current, of people who've chosen another way of thinking and living that makes extreme and healthy longevity possible.

One of the most curious and fascinating accounts of the long-lived patriarchs that I mentioned previously is the account of the prophet Enoch and the way he left this world at the end of his life. I say that he "left this world" rather than "died," because that's what the historical accounts describe. According to the biblical texts, Enoch never died.

Before the book of Enoch was deleted from the official biblical cannon in the 4th century C.E., it held a prominent and revered place in the story of humankind. The book that bears Enoch's name describes how he lived on earth for a total of 365 years, dictating the secrets of creation to a scribe before he left. However, at the end of his time, the texts describe a passing that is not a typical human death.

Rather than expiring with his last breath and his body being returned to the elements, the texts simply say that at the end of his days, "Enoch walked with God: and he was not; for God took him."[2] It is still a topic of controversy in religious and philosophical circles as to precisely what this passage means and what happened to Enoch. I'm sharing the account of his life because it is another description of a life span that exceeds the expectations of modern-day people.

It's after the earthly events described in the texts led to a change in the way humans would live their lives on earth that the accounts of these centuries-long life spans end. From that time until today, a ceiling on age was created, and a limitation was placed upon the length of a human life. Perhaps it's no accident that the biblical account of the limit to a human life parallels the scientific discovery of just such a limit. The biblical parameter is specific. It reads: "My spirit shall not always strive with man, for that he also is flesh: yet his days shall be a hundred and twenty years."[3]

The limit of 120 years described in this ancient biblical passage directly relates to the scientific discovery of a calculator, found within our DNA itself, that determines how many times a cell can divide before it becomes senescent and eventually dies. Each of us has direct access to our cells' calculator, and the discovery that led to a Nobel Prize in medicine is the key to how we can reset the clock that determines the life span of our cells.

TELOMERE SIZE MATTERS

There's a new word that's the buzz of healing and longevity conferences. From television commercials that promise age reversal and renewed sexual vigor to ads suggesting that the medicine of tomorrow is a pill that you can buy on the Internet today, the subject that's suddenly made everyday people sound like DNA experts is the topic of *telomeres.* What telomeres are, and what they do for us, is actually quite simple. What they make possible in our lives, however, borders on the miraculous.

Similar to the way a small plastic cap protects the ends of our shoelaces so that they don't wear out over time, telomeres are special sequences of DNA that protect the ends of our chromosomes as our cells repeatedly divide. For humans the sequence appears as the repeating DNA code TTAGGG, TTAGGG, TTAGGG, and so on. These letters are shorthand for the four possible bases that make up our DNA: cytosine (C), guanine (G), adenine (A), and thymine (T). This sequence is the "stuff" that forms the protective cap seen in Figure 5.1.

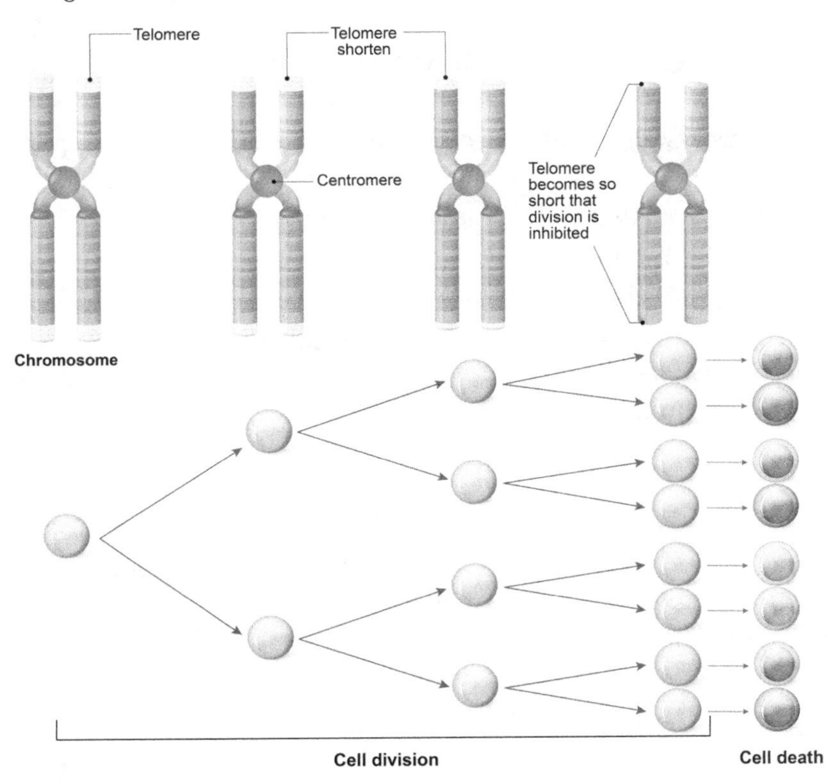

Figure 5.1. This illustration shows how the telomeres shorten with each cell division until they can no longer support the process. Scientists believe that shortening of our telomeres is the biological clock that leads to old age and, eventually, death.

When a cell divides and the chromosomes are copied so two new cells can be created from the original one (replication), the copying mechanism only reads to a certain point along the DNA and then it stops—*before* it actually reaches the end of the strand. This is where the telomeres come in. The telomere is a buffer of additional code that appears *after* the vital information of the chromosome. So when the copying mechanism stops, it stops in the telomeres, where an incomplete copy is harmless, rather than in the DNA information itself. In this way the telomeres take the brunt of the trauma associated with a cell's division. This is nature's program to ensure that our genes are copied completely and that the precious information the cell contains remains whole and intact in its descendants.

If, for some reason, this mechanism did not exist, the copying would stop somewhere in the middle of an important DNA instruction—such as information needed for creating a strong immune system—and the new cell would have an incomplete blueprint to work from. The incomplete copy would show up as a genetic defect that could lead to disease, illness, senescence, and old age. But thanks to the telomeres, this doesn't happen.

With this function in mind, it's clear why the length of our telomeres is so important. As long as they remain long enough to keep the DNA code intact, we have healthy cell division and vital cells that can do what they're made to do.

Key 32: Telomeres are specialized sequences of DNA located at the ends of a chromosome that serve as a buffer to protect the chromosome's genetic information when a cell divides. With each cell division, the telomeres become shorter, until they can no longer protect the vital information of the cell, at which point the cell experiences old age, senescence, and eventually death.

There's a reason that I'm going into this level of detail. Typically the length of our telomeres shortens over the course of our lifetimes. For example, at the time of our birth, our average telomere length is somewhere between 8,000 and 13,000 units (base pairs). As we age, they generally become shorter, and they do so in a predictable way. By the age of 35, the telomeres of a typical adult, living a typical Western lifestyle, are reduced by approximately 29 percent, to about 3,000 units. And when the typical adult reaches the age of 65, that number drops another 50 percent, to approximately 1,500 units. I qualify these statistics with the word *typical* because the length of our telomeres is not fixed. It's not predetermined or "set in stone," as the saying goes.

These statistics describe what happens if we do nothing to support our telomere health. The good news is that *we can* do something. We can do many things. And for this reason, scientists now acknowledge that the speed at which, and the degree to which, our telomeres become shorter depends upon us and a number of factors that we influence through our life choices. These factors include familiar things like diet, exercise, and sleep, as well as detrimental factors such as the use of drugs and alcohol. They also include the less often considered factor of emotional stress that can stem from issues of self-esteem and self-worth.

DISCOVERING THE TIMER INSIDE OUR BIOLOGICAL CLOCK

In 1961 an American scientist named Leonard Hayflick discovered that the number of times telomeres will support a cell as it divides is between 40 and 70 replications. When his discovery is plotted onto a graph of years of age based upon how frequently cells divide, we find what is known as the *Hayflick limit of cell division.* The Hayflick limit predicts the life span of a cell and that limit appears to be the 120 years that we've seen in the previous examples. So whether we're looking at human longevity from a biblical perspective or through the eyes of a biologist, the questions are the same:

- Do we know what causes the limit of 120 years?
- Can we transcend the limit of 120 years?

In light of the new discoveries described in this book, the answer to both of these questions is the same. It's yes!

In 2009 the Nobel Prize in Physiology or Medicine was awarded jointly to three scientists: Elizabeth H. Blackburn and Carol W. Greider, from the University of California at Berkeley, and Jack W. Szostak, from Harvard Medical School. Their award was for the 1984 discovery of an enzyme that is directly linked to the telomeres in our body—*specifically to repairing, rejuvenating, and lengthening the telomeres*. The name of the enzyme itself tells the story. Called *telomerase*, it's associated with the ends of chromosomes, precisely where telomeres themselves are located. The discovery of the purpose of telomerase is best described in the press announcement itself:

> Elizabeth Blackburn and Jack Szostak discovered that a unique DNA sequence in the telomeres protects the chromosomes from degradation. Carol Greider and Elizabeth Blackburn identified telomerase, the enzyme that makes telomere DNA. These discoveries explained how the ends of the chromosomes are protected by the telomeres and that they are built by telomerase. If the telomeres are shortened, cells age. Conversely, if telomerase activity is high, telomere length is maintained, and cellular senescence is delayed.[4]

Key 33: The purpose of the telomerase enzyme in our cells is to repair, rejuvenate, and lengthen the telomeres that determine how long our cells live.

The discovery of telomerase suddenly opened the door to vast new possibilities of healing and longevity. And as often is the case, before humans were involved in exploring the potential of this enzyme, the first studies were conducted on laboratory mice. While a mouse is obviously different from a human biologically, the way a mouse's cells divide and the way those divisions are regulated is the same for them as it is for us. It made sense to test the theories of telomerase, and its role in longevity, on mice before trying them out on human volunteers. The results of the studies were nothing less than astounding.

A 2010 paper published in the prestigious journal *Nature* left no doubt in our minds about what the studies had found. The title of the paper was brief and direct: "Telomerase Reverses the Aging Process." The first sentence of the paper sets the tone for the possibilities that follow, stating, "Premature aging can be reversed by reactivating an enzyme [telomerase] that protects the tips of the chromosomes, a study in mice suggests."[5]

The *Nature* paper described how a group of mice were specially treated in a way that caused them to grow up without telomerase in their bodies while they were developing. The result was that, without the enzyme that could repair their telomeres, the chromosome buffers shortened quickly, and the mice aged faster than they normally would. Not surprisingly, as the mice aged, they developed the same kinds of conditions that we commonly associate with human aging, including diabetes, osteoporosis, and even neurological conditions.

The reason these mice made headlines is because of what happened next. They were also specially treated to have their telomerase enzymes reactivated when they reached adulthood. (This is accomplished by using a specific chemical called 4-OHT.) After the adult mice were treated for one month, they were evaluated. It's the conclusions of their evaluation that were described in the paper.

The lead researcher described the results as "a near Ponce de Leon effect," referencing the Spanish explorer and his legendary quest for the Fountain of Youth. *The age-related conditions of the adult mice were not only halted, they were actually reversed!* "Shriveled testes grew back to normal and the animals regained their fertility," the paper stated. "Other organs, such as the spleen, liver, and intestines, recuperated from their degenerated state. The one-month pulse of telomerase also reversed effects of aging in the brain."[6]

The results of this study have now been replicated and repeated many times and reported in many peer-reviewed scientific journals. Each study approached the aging of cells from a slightly different perspective and tested the role of telomerase, telomeres, and aging in a slightly different way. And as different as the studies are from one another, they're all telling us the same thing. The presence of active telomerase in the body is a key factor in stopping and reversing aging and the deterioration that typically comes with the aging process.

With these studies, for the first time the relationship between telomerase and longevity was confirmed in mice. Since then, the results have been applied to humans as well. While factors beyond the length of our telomeres, such as lifestyle, physical environment, and nutrition, certainly contribute to overall longevity, the correlation between aging and telomere length seems to be undeniable and tells us three things:

1. Longer telomeres are found in people with longer life spans.
2. Telomerase is the enzyme that builds, rejuvenates, and lengthens existing telomeres.
3. Activating the body's telomerase stops further destruction and repairs telomeres that are already damaged.

Telomere length is now accepted as a biological marker—*a measurable sign*—for how long a human can be expected to live.

And what's more, we now know that the marker can be influenced, and intentionally changed, in new and positive ways.

I want to be absolutely clear, however, that simply making our telomeres longer is not a guaranteed prescription for long life. It would make no sense, for example, to lengthen telomeres with expectations of longevity while at the same time indulging in a life of excess that included the chronic use of alcohol and/or recreational drugs, and a diet high in refined carbohydrates, trans fats, and heavily sweetened and fried foods. So while longer telomeres alone do not guarantee long life, researchers have found that only people who have longer telomeres live extended healthy and vital life spans.

As you can imagine, the discovery of the three factors listed previously regarding telomeres and life expectancy has opened the floodgates for new research, an entire new industry of lifestyle coaching, and sales of nutrient and supplement factors designed to lengthen our telomeres, with a promise of long and healthy life. And while some of the products and techniques are based in solid science and actually do what their claims suggest, others are not and do not.

Here's what we know so far.

LIFESTYLE FACTORS FOR LONGER TELOMERES

For those of us trying to keep up with the latest research on what constitutes a healthy lifestyle, tracking the dos and don'ts is enough to make our heads spin. Part of the problem is that the idea of what's good for us is a moving target; the advice is constantly changing. We've seen scientists and medical professionals flip-flop their opinions from the beginning of a year to the end of the same calendar year when it comes to what's good for us and what's not. The thinking regarding chicken eggs and coconut oil are perfect examples.

Chicken eggs: The old idea. In the 1980s, the cholesterol from chicken eggs was thought to contribute to unhealthy levels of cholesterol in the blood and, subsequently, to heart and cardiovascular problems. I remember how eggs were avoided like the plague and how restaurant menus and advertisers went out of their way to alert customers that they were offering egg-free recipes.

Chicken eggs: The new idea. Now the pendulum has swung in the opposite direction, as scientists have recognized that the dietary cholesterol contained in eggs is not the cholesterol that contributes to heart disease or raises cardiovascular risk in healthy people.[7] Instead, the studies show that eggs actually raise "good" cholesterol levels (HDL) and reduce "bad" cholesterol levels (LDL). Suddenly, eggs are the "in" food, recognized as nature's perfect source of protein, iron, healthy fat, and several important vitamins and minerals, and touted as an important element of a healthy diet. And eggs are not the only example of a 180-degree shift in the way we think of certain foods.

Coconut oil: The old idea. Flawed studies that investigated coconut oil in the mid-20th century alerted the general public that this was an oil to be avoided at all costs. For decades I remember ads steering consumers toward other supposedly "healthy" vegetable oils as an alternative to the natural oil of coconuts. Later studies, however, revealed how flawed this thinking was. The original studies of coconut oil were performed on partially hydrogenated coconut oil rather than the natural oil of raw coconuts, and it turns out that it's the hydrogenation process that leads to health-related problems and not coconut oil itself.

By the way, this is true for any oil that undergoes a hydrogenation process, including commonly used oils such as safflower oil, cottonseed oil, corn oil, and soy oil. We now know that canola oil, as well, breaks down into harmful free radicals when heated to temperatures used for cooking above 350 degrees Fahrenheit (176 degrees Celsius).

Coconut oil: The new idea. Now we know that people who live in parts of the world where coconuts are a regular part of the diet actually have a lower incidence of cardiovascular disease than people in countries like the United States, where whole coconuts and coconut oil have been shunned for at least the past two generations. Suddenly raw coconut is recognized not only as a healthy food but as a superfood. Coconut oil in its virgin and extra-virgin forms, as well as extra-virgin olive oil and virgin avocado oil, are now the recommended oils of choice.

Note: Coconut oil is especially healthy because it holds up well under the higher temperatures required for cooking.

In light of the benefits of these two now-recognized aids to good health, perhaps it's not surprising that they are also among the foods that promote longevity and extended telomeres.

As I mentioned previously, there is an entire industry of food and dietary supplements that has emerged touting the ability to heal and lengthen telomeres in recent years. It's impractical for me to describe each and every such product, supplement, and form of exercise in this book. What I can share, however, is that a few key lifestyle factors have been confirmed as necessary for telomere support. In broad categories, these factors include:

- Reducing stress
- Getting regular exercise
- Taking specific supplements

Key 34: Our choices of lifestyle, including specific forms of exercise, specific dietary supplements, and reducing stress in the body, are key strategies documented to successfully slow and even reverse telomere damage and cellular aging.

In the remainder of this chapter, I will identify the factors, techniques, and supplements that I've found through my research and personal experience to have the greatest positive impact on telomeres and aging. And while this list may be used as a guideline, as always, it is important for you to check with your personal health-care provider to see which, if any, may be right for you.

THE VITAMIN AND MINERAL CONNECTION: KEY TELOMERE SUPPORT SUPPLEMENTS

There are a variety of synergistic vitamins and minerals that can contribute to healthy DNA and prevent premature shortening of telomeres. In a study published by the *Journal of Nutrition*, the results showed that men having the longest telomeres also had high concentrations of very specific vitamins and minerals in their blood.[8]

Among the supplements described in the *Journal of Nutrition* study are the following. *Please note:* Some of the following supplements are identified in the smaller units of micrograms (mcg) and some in the larger units of milligrams (mg).

Supplement	Recommended Amount
Vitamin B12	500–1,000 mcg/day
Folate	800 mcg/day
Vitamin C	1,000–3,000 mg/day
Vitamin E	tocotrienols 40 mg/day
Zinc	25–50 mg/day
Magnesium	400–800 mg/day

The entire family of B vitamins is positively linked to longer telomeres. Additional studies have also noted beta-carotene, vitamin A, vitamin D, and the mineral iron as necessary factors for the development and maintenance of DNA and the prevention of premature telomere shortening. A plant-based diet rich in antioxidants and phytonutrients derived from green, leafy vegetables has been shown to be directly related to longer telomere length and healthier DNA.

THE STRESS-TELOMERE CONNECTION

Even before molecular biologist Carol Greider, Ph.D., and her team discovered the enzyme telomerase and identified its ability to reverse the shortening of telomeres, scientists were hot on the track of the role that telomeres themselves play in the aging process. The title of a paper published by the National Academy of Sciences in 2004 sums up the telomere-stress relationship: "Accelerated Telomere Shortening in Response to Life Stress." And although the title sounds a bit complex, the message of the paper is not. In clear and concise terms, the paper leaves no doubt in our minds as to the role of stress in the aging process, stating, "Chronic stress erodes telomeres, impairs DNA replication, and thus accelerates aging."[9]

The key in this consideration of stress is the word *chronic.* This is the kind of stress that goes on and on with no resolution, and this is an important distinction when it comes to the way we think of stress in our lives.

Each of us experiences some form of stress related to something that we either a) need for survival, such as food, water, or medical attention; b) need to produce in an office or work environment; or c) experience as a problem we need to resolve in our personal or work relationships. Interestingly, different people interpret different life events as different kinds of stress.

We've all heard of *constructive stress,* for example. This is the kind of stress that creative people often feel in response to a deadline or the pressure of creating or producing something to meet a specific goal or need. An artist under the stress of creating paintings for a gallery opening, an office worker creating the financial reports that are due at the end of the quarter, an author writing to meet a publishing deadline, and a scientist solving the problem of how to save precious power in an orbiting space capsule that is desperately low on energy (as in the *Apollo 13* story) are all examples of situations that lead to constructive stress.

Each of these stress situations has a beginning that can be identified and an end that can be recognized and met. In these situations the primary stress hormones of adrenaline, norepinephrine, and cortisol become the fuel to kick creativity and problem solving into high gear. As new solutions appear, the goal appears to be within reach, the sense of being stressed eases, and the stress hormones dissipate.

Herein lies the key to constructive stress: It's temporary and the effect of the stress chemistry is typically short-lived. As we near our goal, we may say that we "see the light at the end of the tunnel." The fact that there is a light, that we feel ourselves getting closer to it, and that the heightened effects of stress upon the body are short-lived is what makes our creative stress a positive experience.

This is much different from the way we feel when we're in a demanding or difficult situation that seems to have no light at the end of the tunnel, no end to the stress. And when a situation seems especially hopeless, sometimes we can't even see the tunnel itself, which would make movement toward the light at its end possible. Working for a big company where we feel like we're just a number on a spreadsheet in some executive's office, for example, can create this kind of stress. In such a situation, it's obvious that no matter how hard we work each day or how innovative we are while we're working, the current conditions that are frustrating, unhealthy, or even hurtful probably aren't going to change. No matter what we do, how hard we work, or how good our work is, our frustration remains unresolved. It's under conditions like these that stress can become chronic and harmful.

The sense of helplessness from this kind of experience triggers a chemical reaction in our body that's called the *fight-or-flight response*. In this mode our biological impulse for survival kicks in and triggers higher levels of the stress hormones mentioned previously to prepare us to do one or the other: either fight our way to safety or run like hell until we've escaped the threat. And while this kind of response would have served us if we were fleeing from a saber-toothed tiger at the end of the last ice age, in the office or home environment today it gives us a completely different kind of experience.

ANCIENT STRESS IN THE MODERN WORLD

When our ancestors successfully escaped a saber-toothed tiger and they were catching their breath while resting behind a big rock, for at least a moment the immediate trigger for their stress was resolved. The levels of stress hormones in their bodies would begin to decline as their heartbeats slowed and returned to normal. And after a few hours of relative safety, they would have metabolized the high levels of these hormones from their bodies.

In this type of dangerous situation, our stress hormones serve us well by being available when we need them in high amounts for brief periods of time. But this scenario is not what typically happens in our lives today. In the modern world, we're usually not physically chased by another animal that is threatening our lives. Instead, our stress often comes from finding ourselves in situations where we feel trapped, vulnerable, and helpless. And in these cases, the resolution is not as cut-and-dried as successfully eluding a hungry animal.

This is where the problem arises. On the one hand, the body is revved up with the primal chemicals to run, hide, or fight, while on the other hand, we usually can do none of these. It's like we're in a car with one foot pressing on the gas, ready to go, and the other foot on the brakes at the same time. The engine is revved to the max and we're not moving.

When the source of our stress is the secure job that pays our bills yet we hate with a passion, or it's the 15-year-long relationship that we feel trapped in but that provides security for us and our children, we can't run and we can't hide. At least not in the way our ancestors did when they found refuge behind a rock.

In the modern world, where is our rock to hide behind? If we haven't discovered a way to feel safe and relieve the stress of daily life, studies tell us that the unresolved stress will begin showing up in ways that negatively impact our telomeres. And while the science that describes this relationship is clear, the effects of the stress are obvious even without the science.

When someone we know is stuck in the throes of unresolved emotional turmoil—as they would be during a long and drawn-out divorce, or if they are unable to make a decision whether to continue a job or relationship or to end it, for example—we see the toll the stress takes on them. We see it in their aging bodies and in their aged faces. They look older than their years and typically start to have health problems that would normally not be seen until years, even decades, later in their lives. When people are chronically stressed, their immune systems are often not prepared for the cold and flu that invariably sweeps through the office or the classroom each year.

These are the people who must use all of the sick days they have and then need even more. And ultimately, these are the people who succumb to stress as it steals from them the very thing they cherish most: life itself. In the presence of long-term, chronic, unresolved stress, their bodies can only hold up for so long.

Staying healthy all comes down to our ability to give the body the environment it needs to do what it is designed to do—heal—and do so at the most fundamental level of DNA itself.

Key 35: It's the *unresolved* stress in our lives that erodes our telomeres and steals from us the very thing we cherish most: life itself.

DIVERSIONS: HEALTHY AND NOT

Our instinct is typically to do whatever we can to avoid facing situations where we feel there's no resolution. So we create diversions in our lives that take our attention away from the issue, or issues. We can engage in healthy diversions, such as yoga, meditation, individual or group sports, art, or music, to channel our stress. Too often, however, the diversions we opt for are less healthy choices, such as eating when we're not hungry, using drugs or alcohol to numb our unpleasant emotions, or resorting to online video games or even video relationships in place of in-person interactions. When we choose these activities, it's often a way of diverting our focus from feeling the emotions associated with the stress.

When we become addicted to the release of the chemicals in our bodies that our diversions create—such as serotonin and oxytocin, which improve our mood—the activities that produce them will become our chronic go-to escapes. We'll eventually feel addicted to them. And unless we find a way to resolve the underlying stress itself, over time such diversions can replace our friendships, jobs, family, and other primary relationships as our sources of good feelings.

This is what the paper from the National Academy of Sciences is saying to us. And it's also telling us precisely how chronic stress is harming us: by shortening the telomeres that protect the code of life that lives in each cell of our bodies. The good news is that the same science that tells us chronic stress hurts us also tells us how we can resolve that stress.

EXERCISE

Discovering and Resolving Your Unresolved Stress

The science is clear: Unresolved stress can shorten the telomeres that are crucial to your health, healing, and longevity. I've created a concise template to help you to identify this kind of stress in your life. I've found this simple template to be particularly useful when you feel a sense that something is bothering you, yet have been unable to clearly identify what that something is. I invite you to take this opportunity to clarify any stressors that you may be experiencing in your life at this moment. You'll need paper and pen for this exercise.

The Technique: Using single words or brief phrases, please write down your answers to the following three questions as honestly as you can.

Question 1: What are some of the sources of your unresolved stress? Please identify, as honestly as you can, any relationships, conditions, or situations in your life that create a lingering sense of anxiety and frustration or a deep, emotional, "gut" response of uncertainty when you think of them. Make a list, reserving a few lines of space below each source you identify.

Question 2: What is your typical response to your stress? In the space below each source of stress, I invite you to identify diversions you typically rely upon. Finish the following sentence: "When this situation triggers feelings of anxiety, frustration, or any other deep emotion that is uncomfortable, to feel better I typically . . ."

Question 3: What would you like to be your new and more thoughtful response to your stressors? If you'd like to replace your current diversions for the sources of stress you've identified with new and more thoughtful responses, please follow the following strategy. It begins with making a familiar shift to ask your heart's wisdom for guidance. You'll recognize the steps here for creating heart focus as similar to the Quick Coherence® Technique described in Chapter 5.

- **Step 1: Establish a heart focus.** Allow your awareness to move from your mind to the area of your heart.
- **Step 2: Slow your breathing.** Begin to breathe a little slower than usual, taking approximately five to six seconds to inhale and the same to exhale.
- **Step 3: Access your deep intuition.** As you continue to breathe and hold your awareness in your heart, silently ask your question from within.
- **Step 4: Listen/feel.** Listen for the answer. When you've got one, write your heart's wisdom below. Finish this sentence: "Examples of more thoughtful responses to my unresolved life stress include . . ."

The purpose of this exercise is twofold. Use it to:

- Develop an awareness of the diversions that you turn to,

both consciously and unconsciously, when you're faced with stressful situations that seem to have no resolution.

- Replace any diversions that may not be serving you with new and healthier responses to the stressors in your life. The key to the power of this exercise is that although we may not always be able to change the situation immediately, we can immediately shift our response to it.

As you complete the exercise above, I invite you to bear in mind that there is no correct or incorrect way of receiving your heart's wisdom. Each of us is born with our own unique code that allows us to access our heart's wisdom and apply it in our lives. The secret to the code is to know what works best for you.

Key 36: Through our heart's wisdom, we can ask for and receive insights into healthy alternatives to the unhealthy diversions in our life.

WE AFFIRM OR DENY OUR LIVES IN EACH MINUTE OF EVERY DAY

If we embrace a way of living that constantly fills our "tank of life," as Michael described it, then we constantly replenish the vitality of our cells. When we do so, our telomeres continue to heal, grow, and divide in a way that reflects that vitality. While the instructions for this kind of thinking and living are simple, to implement them is where the real workshop of life begins. It takes courage. It takes discipline. And it requires a choice that we must make in each moment of every day of our lives.

Through the choices we make in each moment of every day—the food that we choose to nourish our bodies, the movements that we choose to stimulate our bodies, the words that we choose to express our thoughts and experiences, and the beliefs that we harbor about ourselves and other people—we either affirm or deny life in our bodies. As we embrace the fact of this reality, our choice becomes simple. It's based in the mindful decision to choose life in every word, through each meal, and in every interaction that we have with ourselves, with other people, and with the world at large. This key to longevity is no secret to students of the mystery traditions, such as the ancient Essenes, a religious sect that flourished from the 2nd century B.C.E. until the 1st century C.E. throughout the land that today includes parts of Palestine, Jordan, and Israel.[10] Perhaps the most widely recognized Essene today is the New Testament master Jesus of Nazareth.

Jesus described the choice of life that we make each day to his students in a way that made sense to them, in the vernacular of their day. When they asked him how they could heal their bodies, Jesus replied in a way that was direct, simple, and eloquent: "If you eat living food, the same will quicken you, but if you kill your food, the dead food will kill you also. For life comes only from life, and from death comes always death. For everything which kills your foods kills your bodies also."[11]

The best science of the modern world is telling us that these direct, powerful, and eloquent words are just as true today as they were 2,000 years ago. When we eat foods that are heavily processed, overcooked, or laden with preservatives, we're eating food in which the vitality of the enzymes and the life that nourishes us has been killed.

An accepted definition of *food* is "any nutritious substance that people or animals eat or drink, or that plants absorb, in order to maintain life and growth."[12] From this definition we can see that what are commonly called processed foods are actually not foods at all. While they may fill the space in our stomachs and ease our hunger, because the components that go into these fast meals are already dead when they're packaged, they cannot give life to our bodies when we eat them.

So it comes as no surprise that popular diets consisting of fast foods laden with hydrogenated oils, processed ingredients, preservatives, and artificial colors and flavoring are implicated in the epidemic of illnesses that are sweeping the modern world, such as diabetes, dementia, and various cancers. It makes perfect sense when we consider that we may be relying upon nonfoods to give nutrition to our bodies.

And as powerful as this realization is for our diet, the same principle applies to the choices we make beyond what we eat. It applies to what we believe about ourselves and other people, to our relationships, and to our self-esteem. Each of these is a spiritual food for our minds and hearts that nourishes us in some way. To build upon the wisdom of the Essene teachings with these ideas in mind, we can take them one step further and say, "For everything which kills your sense of value, self-worth, and self-esteem kills your body also."

Clearly, the quality of our emotional, psychological, and spiritual nutrition is just as important as the nutrition from the physical food that we eat.

The key here is that the quality of each of the telomere-healing experiences in our lives is based upon the choices we make. Sometimes we make our choices consciously and with intention. Sometimes our choices are unconscious. Regardless, they are always our choices. *The key to longevity is to turn our conscious choices into subconscious habits.* When we do this, we no longer need to stop and ponder what we'll eat for lunch or how we'll respond to a lovers' quarrel. The reason is that we've already chosen.

I learned this lesson for myself early in my life and it has become the framework for the majority of choices I make each and every day. Each day, with each meal I choose, with each friendship, partnership, and relationship I'm involved in, and when I find myself criticizing another person or myself for something I've said or done, I ask the same question: *Is this the best that I have to offer in this moment?* The answer to this question in the moment tells me what my possibilities are—and it's at that point that I must make a choice that affirms or denies the very life in my body.

With the relationships between nutrition, beliefs, stress, and telomeres in mind, it's clear that longevity is actually less about trying to see how long we can live and more the product of a choice that we make in each moment of every day. This choice is what my conversation with my friend Michael was all about. It's the difference between thinking of ourselves as finite vessels of limited potential and thinking of ourselves as infinite vessels of unlimited potential. That's the difference that would make it possible to become a father at the age of 500 years.

Key 37: In each moment of every day, we make the choices that affirm—or deny—life in our bodies.

TIME, LIFE, AND THE AGING CLOCK

During the years that I was leading groups in Tibet, I noticed a phenomenon that is rarely discussed in textbooks or travel documentaries. It's simply that the Tibetan monks and nuns typically don't track their age. The first time I asked a Tibetan monk his age, his initial response was to begin laughing. He wasn't laughing at my poor Tibetan language skills; rather, he was laughing at the question I'd just asked. He didn't believe it was a serious question, because in his way of thinking, age doesn't have the same meaning that we give to it in our culture.

After the monk realized I'd asked a serious question, he was more than happy to answer me. His delay in responding wasn't because his age was some kind of secret. It was because he simply didn't know. Tracking the passing of each year in his life wasn't important to him.

While the monks celebrate their birthdays, they don't actually track their age. They celebrate the successful completion of another revolution around the sun rather than counting how many years have passed since they were born. From the previous sections, we know that the consequences of this way of thinking are clearly positive. If the expectation is that our quality of life declines with each passing year, and that counting the years we've lived affirms our aging, then it makes sense that the monks would want to avoid tracking their age.

My monk friend certainly knew the year he was born. And with that date in mind, he began his answer to me by first asking me a question in turn. "What is the year now, today?" he asked. When I replied that the year was 2008, he nodded his understanding. He looked at his open palm and began scribbling invisible numbers with his index finger. He was calculating the difference between the year I'd just stated and the year that he was born. Quickly he looked up at me with a big grin and proudly told me that he was born in 1915. By his calculation he'd been in the world for 93 years.

His answer more than surprised me. If I was being paid to estimate the age of this man, from the color and tightness of his skin, the sparkle in his eyes and the bounce in his every step, I would have guessed that he was maybe in his mid-60s or possibly his early 70s. But there is no way that I would have placed him in his 90s! Just as the nun at the monastery had demonstrated, this monk had shown me that the number of years we're in this world and the condition of our bodies are not necessarily linked in the way I'd been taught to expect.

The lesson I learned from the monk was about time.

AGING DOESN'T MEAN OLD AGE!

If we set the timer on our cell phone for 60 minutes, at the end of the 60 minutes, we will all have been on the earth for one hour more than we were when we set the timer. The 60 minutes marks the chronological time that's lapsed since we set the timer. And while we will have clearly lived through each and every one of those 60 minutes, the question will be, *how* did we live them? While our cells have lived and metabolized during that hour, did they also heal and rejuvenate during that time? And perhaps more importantly, did we give our cells the environment to heal and rejuvenate? Our answer to this question is the difference between longevity and old age.

The very nature of this question goes back to the philosophy at the beginning of this chapter. Do we believe that we begin to die from the moment we're born, or do we accept that the moment of our birth triggers the healing process that is natural and inherent in our bodies' existence? To make this more personal, do you believe that from the moment of your birth you've been healing and rejuvenating?

The key to what the Tibetan monk demonstrated in his answer to me is that he didn't tell me he was "93 years old." He didn't claim to have used up 93 years of his finite tank of life. What he told me was simply that 93 years had passed since he came into this world. In other words, he stated the fact of his longevity without affirming the consequences of his age. This subtle way of honestly acknowledging our time on earth has powerful implications when it comes to the age clock in our cells. It's the key to the longevity and quality of life that I found at first with the monks and nuns in Tibet.

Since I've learned to recognize this philosophy, I've now seen it in many indigenous traditions that are less influenced than the Western world by thoughts of life, death, and longevity.

One of my lifelong passions has been the study of people living to advanced ages who have done so in a healthy way. My search has been to discover the common denominators that are shared by the oldest people in the world. When the monks told me that there are yogis who are 600 years of age, as remarkable as the claim was, I actually felt I had no reason to doubt them.

The new discoveries of modern science certainly suggest that such advanced ages are possible, and ancient writing tells us that humans have reached these ages and survived even beyond them!

What's most important to me about these stories is that when these people finally arrive at the end of their centuries-long lives, they don't fit the contemporary idea of what a person of such an age should look like. What I mean is that they don't look like the image of a shriveled body of wrinkled skin hanging on a fragile skeleton that we often associate with advanced age. Just the opposite. These people, like the nun I met in Tibet in 2008, have bright and focused eyes, healthy and supple skin, and extremely active lives. They're vital, fully enabled, and fully capacitated humans, enjoying their lives to the fullest and contributing to their families and communities to the end of their days.

And while we don't have documentation for the yogis that my guide described, we do have documentation for one man who lived to a near-biblical age in relatively recent times. One of the most fascinating, extreme, and best-documented examples of reaching a biblically old age is a man who served in the Chinese military and was honored by the military for his 100th, 150th, and 200th birthdays, Li Ching-Yuen.

THE MYSTERY OF LI CHING-YUEN

Li Ching-Yuen was a Chinese martial artist and qigong master who lived on a diet of high-altitude herbs, served in the Chinese military, and died at the advanced age of 256 years. The detailed military records of the Chinese army indicate that Li was born in Sichuan, China, in 1677. His entry into military service as a tactical advisor in 1749 is well documented, as well as is his retirement 25 years later at the age of 97. In his retirement, he resumed the simple rural lifestyle that he had led before his military service. I believe that this choice of lifestyle was one of the keys to his longevity. He returned to the high mountains of China's Sichuan

Province to grow, harvest, and sustain a diet of medicinal herbs just as he'd done before his military service.

In recognition of his distinguished military career, Li received a letter of acknowledgment for his service. Accompanying the letter was an additional document of good wishes congratulating him on his 100th birthday. The year was 1777. The military acknowledged him again in 1827 on his 150th birthday, and yet again in 1877 on his 200th birthday. This mysterious man of longevity is reported to have died in 1933. I have to say that he is "reported" dead because in the rural setting of his hometown, his body was never seen and his family never buried him. According to his wife, he simply died while he was in nature.[13]

Figure 5.2. A rare photograph of Li Ching-Yuen taken in Sichuan in 1927, when his documented age was 250. Records from his military service indicate that he was born in 1677 and died in 1933. At the time of his death, he is believed to have been 256 years of age. Source: Public domain, People's Republic of China / Wikipedia

In 1933 both *Time* magazine and *The New York Times* published articles about Li Ching-Yuen containing interviews with descendants from the village where he had grown up.[14] The articles described the memories of adults who had known Li in their childhood, but their accounts were shared by the grandchildren of these adults. At the time of his death, Li was credited with 180 children from 14 marriages. When he was asked to what he attributed his longevity, Li said he believed that the secret to his long life was to "have a quiet heart."[15] In light of the new discoveries regarding the effects of a heart-focused life, Li's words make perfect sense.

The reason I'm sharing this account and my experience with the Tibetan nun who was 120 years of age when we met is not about their age specifically. While the advanced ages of these people are clearly impressive, really it's the exceptional condition of their bodies at the time of their age that's the point of this discussion. Both people illustrate an exception to the thinking that "we begin to die the moment we're born." In fact, they seem to support the possibility that I shared with my friend Michael.

It's only through a process of continuous healing—a rejuvenation that begins at the DNA level of life itself—that it's possible to achieve such remarkably old ages.

I would have loved to have a conversation with Li Ching-Yuen before he left this world. I would have asked him the same questions that come to anyone's mind when we hear of life spans that challenge our belief systems: about diet, exercise, and lifestyle. Unfortunately, Li passed two decades before I was born, so I missed my opportunity.

LONGEVITY: THE COMMON THREAD

In 2008, the Associated Press carried the story of Mariam Amash, an Arab-Israeli woman from the village of Jisr az-Zarka in northern Israel. She was flagged at a security checkpoint that year, reportedly because her identification papers had expired. She was told that she needed to go to the local authorities to have her documents updated and renewed. That's when she made worldwide headlines. She was issued new documents that showed her date of birth. Mariam's confirmed birthdate was 1888, meaning she was 120 years of age at the time her story made headlines![16]

When she was asked what she attributed her health and long life to, Mariam didn't need to think long before she answered: It was love. She felt that the love she had for her family—the love she *felt* for her children, her grandchildren, her great-grandchildren, and her great-great-grandchildren—was what kept her alive for so many years. She felt important in their lives. She cared for them. She cooked for them. She counseled them in life. And each of these experiences contributed to one overarching positive factor: Mariam felt needed. She felt that she was contributing to the lives of the people she loved in a way that they needed. And it was this feeling that drew her to a full life each and every day.

In 2012, one of her grandsons reported to the press that Mariam had not been feeling well. She was taken to Israel's renowned Hillel Yaffe Medical Center, in the city of Hadera, for observation and treatment. Only three days later, without a prolonged illness, Mariam passed away quietly, surrounded by her family. At the time of her death she was 124 years of age and had 10 children and approximately 300 descendants. I'm sharing this story because, similar to Li Ching-Yuen, Mariam led a vital and healthy life up until the very end.

When we think about the examples of the three people that I've described, Li Ching-Yuen, Mariam Amash, and the Tibetan nun, a single through-line becomes immediately obvious. All three of these people attributed their longevity to positive, heart-based experiences. Knowing this, it should come as no surprise that the positive experiences of a peaceful heart and of being loved—of feeling loved and needed—would impact these people's bodies in a powerful and positive way. It's the new science-based discoveries, however, that give us the specifics. And when we understand the precise relationship between our perception of life's experiences and longevity, we also discover how to awaken that ability consciously in our own lives.

Every organ in the human body, it is now documented, has the ability to regenerate and heal, including the organs that we've been told in the past are incapable of doing so. Heart tissue, brain tissue, spinal-cord tissue, pancreatic tissue, and even nerve connections are all now documented as having the ability to repair themselves and heal damage they've sustained, and to do so using the body's own healing mechanisms. The discovery of telomerase tells us why this universal healing is possible.

The key is that we must create the right conditions—the right environment inside and outside our bodies—to trigger such healing. These conditions can include the physical environment of our surroundings, the chemical environment of our blood and cells, and the emotional environment that triggers our heart and brain functions. This discovery has opened the door to a new reality in biology and a new way of thinking about life, which begins with the discovery of cells that can live forever—the first immortal cells.

THE FIRST IMMORTAL CELLS

When the 2009 Nobel Prize was awarded for the discovery of telomerase, it was like the last missing piece of a puzzle had dropped into place for longevity research. Biology textbooks have historically shown an illustration similar to Figure 5.1, where telomeres become shorter and shorter each time a cell divides. And because the number of times cells can divide was thought to be limited (known as the Hayflick limit), cells were said to be *mortal*. It was believed they had a life span that could be calculated, and the number of times the cell could divide could be predicted.

With the discovery of telomerase, however, and its ability to extend the length of telomeres and the life of the cell, a new class of *immortal cells* had to be created. The reason for this name is that the cells are not bound by the Hayflick limit. In theory, as long as the telomeres continue to be healed and replaced, a cell can continue to live, grow, and thrive. And in theory, this process could happen indefinitely, making the cell immortal. While the idea of immortal cells may sound like science fiction, the reality is that they already exist. And the fact of their existence is not a recent achievement either. The first immortal cells were discovered in 1951. And the shocking truth is that those cells are still alive and reproducing themselves in laboratories today, some 65 years after they were first recognized.

In 1951, a doctor at Johns Hopkins hospital created a cell culture from tissue taken from a young woman who had cervical cancer. In her particular instance, as with many cancers, the body's naturally programmed cell death that normally kills defective cells before they become a problem—*apoptosis*—was not working. Rather than killing off the cells that had not divided properly, her body was sending a signal to do just the opposite. It was producing telomerase to keep all of her cells alive and reproducing, including the defective ones. This is why the doctor made a laboratory culture from a sample of the woman's cells. He wanted to understand why the unhealthy cells continued to live and reproduce in this manner.

The woman's name was Henrietta Lacks, and her cells continue to reproduce as tissue cultures today. The original culture that the doctor created in 1951 keeps perpetuating itself, and the cells that it produces are studied throughout the world in classrooms and medical research laboratories. They're known as the HeLa cell line, to honor the name of their donor. In theory the HeLa cells may live forever.

In Henrietta's case, something unknown triggered a blanket release of telomerase in her body in 1951. It could have been an environmental toxin. It could have been her body reacting to an additive or a preservative that was used in mid-20th century products that no longer exist. It could have been a concentration of heavy metals in her environment. What's important here is the fact that Henrietta Lacks's cells are still alive and will continue to reproduce as long as a constant supply of telomerase is present.

ARE WE REALLY READY FOR IMMORTAL CELLS?

The existence of Henrietta Lacks's perpetually dividing cells has taken the idea of cell immortality from a textbook theory to a physical reality. The question is no longer whether or not it's possible to produce ever-living cells. Now the question is whether or not this immortality can be safely induced in a healthy human through diet, exercise, nutrition, and supplements. And if the answer is yes, then the question that follows is more of a philosophical one: Are we really ready for immortality and what it means in our lives? Are we emotionally prepared to live extended lives where we outlive everything that's familiar and everyone we love? The answer to this question is something that scientists are now seriously considering. They need to, because it appears that we'll need those answers sooner rather than later.

Throughout recorded human history, and possibly even before, our lives followed an unspoken pattern when it comes to relationships, career, and family. Historically, the pattern has gone something like this: Sometime after the end of childhood, which is considered to be as early as puberty in some societies, we organize our lives by working to define a career path. We seek out a life partner, and some people begin to create families. As we have our own children and guide them through their formative years, the natural order has been for us to live a full life—and if we are fortunate enough, to become grandparents—and then, due to the complications of aging, to pass on, leaving the fruits of our lives for the next generation.

Our society is geared toward this progression, which is commonly known as *the natural order of life*. The structure of our careers, retirement, Social Security accounts, and medical insurance plans are all based upon statistics projecting how far into the future we are expected to live and how we will need to draw upon their support. These statistics reflect the average among our peers and within the natural order. Today, expectations are changing. As technology, hygiene, and workplace safety have improved over the years, life spans have increased, and the statistics are reflecting that change.

For example, in 1930 the average life expectancy for a man was 58 years and for a woman it was 62 years. The difference between these ages is commonly attributed to the hazards of factories, underground mines, and the toll of war, which affect men differently than they do women, as well as the tendency for cardiovascular problems to affect men earlier in life.

Interestingly, the retirement age in 1930 was 65, meaning the expectation was that most people would work their entire lifetimes, without ever experiencing the benefits of an official retirement or receiving a payout from a social program. Fortunately, improved standards of work and living have changed these numbers significantly. According to the U.S. Social Security Administration, in 1990 if a man survived the stress and perils of life and career to live until age 65 and then retired, he could expect to live an additional 15.3 years following his retirement. For women, the statistics are even better. In 1990 a woman could expect to live for an average of 19.6 years after her retirement, 4.3 years longer than her male counterpart.[17] Even in light of these new statistics, for the most part the natural order of life has remained intact in developed nations.

When it comes to families, a similar progression is assumed. The expectation is that parents provide for their children while they're maturing, and when the parents die, they leave their material wealth and the fruits of their lives for their children to enjoy. Our partnerships and marriages are based upon this same model and progression. When we make a lifetime commitment in marriage, for example, we assume that we're committing to a life span within the historic range of life expectancy. The possibility of immortality, or even a lifetime that's extended by 100 years or so, changes all of this. Honestly, how many people would commit to a lifetime with a single partner if they knew in advance they would live 200 years? Or how about 500 years? Or achieve immortality?

And while it's possible to adjust the nuts and bolts of the material world—such as finances, insurance, and jobs—to accommodate longer lives, perhaps the greatest challenge for someone living a lifetime measured in centuries is the emotional toll of the losses that they would experience over the course of their long life. In a multicentury lifetime, there is a very real possibility that the person living that long would lose everything they've known and everyone they've loved. They would experience the loss of friends, family, partners, and lovers, and each loss would need to be acknowledged and healed in some way. This would be especially hard when it comes to parents and their children. *Psychology Today* describes the emotional impact of a parent outliving their children:

> Producing greater stress than dealing with the death of a parent or spouse, a child's death is especially traumatic because it is often unexpected as well as being in violation of the usual order of things in which the child is expected to bury the parent. The emotional blow associated with child loss can lead to a wide range of psychological and physiological problems including depression, anxiety, cognitive and physical symptoms linked to stress, marital problems, increased risk for suicide, pain, and guilt.[18]

In addition to the loss of loved ones, a person living a multicentury life would also experience the loss of neighborhoods, communities, and entire ways of life and living, as the world would continue to grow and evolve, changing dramatically during an extended life span. It's this very scenario that has long concerned scientists when they think of astronauts journeying on multidecade missions to other worlds, when the phenomenon of time dilation predicted in Einstein's equations becomes a very real factor. Space travelers' families and friends on earth would continue to age at the normal rate, while those aboard a spacecraft, due to the speed they would be traveling, would age more slowly than their earthly counterparts. (This is one of the implications of Einstein's $E = mc^2$ formula.) Assuming that they survived their decades-long mission, when they returned to Earth, they would be much younger—depending upon how long they were away and how fast they were traveling—than the people they left behind.

While none of the scenarios I'm mentioning here is necessarily a showstopper when it comes to extended life spans, they all offer a hint of what's involved with the experience of longevity in a way that goes beyond simply keeping cells alive. It all comes down to our perceptions and how we feel about the changing world around us.

I experienced a taste of this for myself with my grandfather before his passing.

KEEPING UP WITH THE WORLD

As I mentioned previously, my father left our family when I was 10. My mother's father became more of a father to me after my dad's departure, and I was closer to him than I was to my biological father. Though my grandfather and I had very different world views, he was always open to new ideas, willing to listen to my concerns, and happy to share his wisdom when I asked—or when I needed it most. While I didn't know it would be the last week of his life, I was with my grandfather the week he died. He had just turned 96 and we had a small family celebration to acknowledge the experiences of his lifetime.

As the celebration was winding down, I pulled my grandfather over to a quiet table and asked him to tell me about his 96 years of living and what it meant to him. Having left the noise of the party in another room, he began by taking a deep breath as he raised his eyebrows and rolled his eyes at the magnitude of what I'd just asked. "There was a time when the world made sense to me," he said. He then described how he had understood the world and how things worked, and prided himself in his masterful ability to fix things when they needed to be fixed. This included repairing the engine on his own car as well as the cars of his friends and relatives, maintaining the coal-burning furnace in his family's basement through the tough Missouri winters, and always being able to work for what he had, even during the Great Depression of 1929, and pay cash for his home and furniture without ever receiving any handouts from anyone. This time he was talking about, when the world made sense, was in the 20th century, just after World War I.

"Then something changed," he said, "and the world didn't make sense to me anymore. I couldn't keep up with the changes." Grandpa could never put his finger on any one item that was responsible for the changes that left him feeling like an outsider. "It was everything," he said. "Everything changed!" Just after World War II, the fruits of wartime technology began to trickle down into everyday life. From jet planes and telecommunication systems—like fax machines—to entirely new kinds of medicine and entirely new industries, the gadgets and ways of life that emerged after World War II worked on principles that my grandfather simply didn't understand.

In addition to the flood of new technology, the world was full of new nations as well. Many of them had not existed before the war. (These were nations such as Israel, Jordan, Pakistan, Iraq, and Nepal.) Grandpa could never understand how one day a nation didn't exist, and then the next day, with the stroke of a pen, suddenly it did. All of this left my grandfather with a sense that he didn't fit into the world any longer—that he didn't belong. At the age of 96, he couldn't reconcile the changes in the world within the context of his own life.

I wasn't with my grandfather when he died later that week. I received a phone call while I was at work letting me know that after his lunch, Grandpa had taken a nap in his favorite chair while watching daytime TV, with the brim of his University of Missouri ball cap pulled over his face, and he never woke up. His was a peaceful transition, and I've always been grateful for that—and that I asked him my questions about his life when I did. Sadly, my grandfather died feeling like a stranger in the world he grew up in. I think of him often and what a century of change meant to him, and I wonder what it would mean to have an even greater amount, maybe two centuries of change or even more, to reconcile within a single lifetime. The good news is that the same science that makes longevity and immortality possible has also come full circle with the knowledge to reconcile what such change means in our lives.

EMBRACING BIG CHANGE IN A HEALTHY WAY

Perhaps it's no accident that the components driving change in our world today—such as the technology and the discoveries that led to the development of immortal cells and to recognizing the power of heart-brain coherence—have all advanced at the pace that they have. As the discoveries emerge in the same period of time, it becomes clear that each needs what the others offer in order to be useful in our lives.

In Chapter 3 I described the discovery of the conversation between the heart and the brain (coherence) and the many benefits that are available to us as we optimize this conversation. In addition to the extraordinary abilities of deep intuition, super learning, precognition, the triggering of a powerful immune system, and the release of the life-affirming enzyme telomerase that I mentioned, there is an additional benefit that heart-brain communication makes possible as well. It's called *resilience,* and it is nature's way of helping us embrace big change in a healthy way.

In recent years, scientists have discovered that by increasing our resilience to life's challenges, we actually reduce the stress that those challenges can create in our lives. In other words, as we strengthen the conditions within our mind-body-emotion system, we shift the way we feel about our life challenges—our perceptions—and do so in a healthy way. This is possible even though the circumstances that are the source of the challenge may not have changed. It's this kind of resilience that will become the key to healing the emotional hurts described previously that we experience through extended life spans. The beauty of increasing our resilience in life is that we can do it at any age and at any time in our lives.

Key 38: Heart-brain resilience is the key to emotional healing from the loss of family and loved ones that comes with extended life spans.

A NEW RESILIENCE

Whether we're talking about one person or an entire planet of people, the conventional thinking when it comes to resilience is that it's an inner quality that allows us to recover from a challenging event that's happened in our lives, such as the loss of a loved one, job, or relationship. The American Psychological Association defines this kind of resilience as "the process of adapting well in the face of adversity" and "bouncing back from difficult experiences."[19] As well as this traditional definition fits the conditions described on the evening news, and as much as the thinking underlying it makes sense, there's another kind of resilience. It's a new form of *expanded resilience* that's seldom discussed, yet when we learn that it exists, it makes perfect sense.

The Stockholm Resilience Centre describes resilience as the capacity to "continually change and adapt yet remain within critical thresholds."[20] It's the theme reflected in this second definition that best illustrates the kind of resilience we need in order to embrace the changes experienced within an extended life span. We're talking about a way of thinking and living that gives us the flexibility to *continuously change and adapt* to new challenges, new conditions, and new ways of thinking and living, rather than having to bounce back from one loss after another. And it's this form of resilience that's the key to healing unresolved stress. If we think of our personal resilience as the combined force of the emotional, physical, and psychological "batteries" that power us through life's challenges, then expanded resilience is the juice that keeps our batteries continuously charged.

It all begins with the resilience that we create within the heart itself. One way of determining our level of resilience is to measure the peaks and valleys of our heart rhythms.

A DEEPER RESILIENCE FROM WITHIN

While most people are familiar with the graph of our heart rhythms that a doctor scrutinizes at our yearly medical exam, we may not be fully aware of everything the graph is showing. In addition to telling us information about the overall condition of the heart, these rhythms can also tell us about the health of the nervous system. The graph the doctor is looking at is probably an ECG, or electrocardiogram. The ECG measures the electrical output of the heart—the electrical impulses that the heart creates and sends throughout the body.

While the study and interpretation of heart rhythms could fill an entire book, I just want to focus on one thing about them here. There's an aspect of the heart rhythm that's the key to creating resilience. When looking at the peaks and valleys of an ECG chart, even an untrained eye can clearly see that there are repeating patterns of large spikes created by each heartbeat (see Figure 5.3).

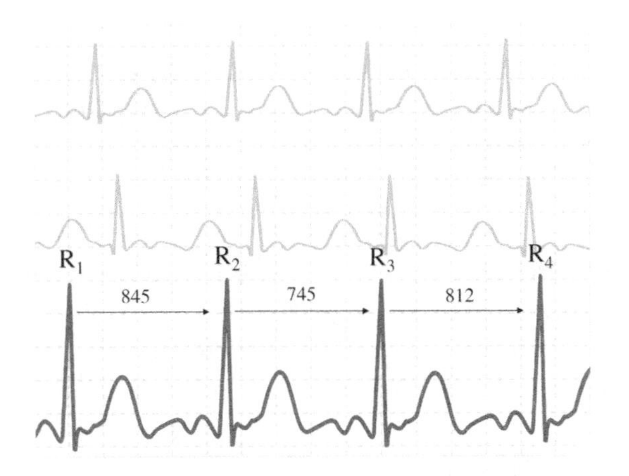

Figure 5.3. A portion of a typical ECG, showing the cyclic peaks and dips of a typical heartbeat. The distance from the peak of one R wave (R1) to the next (R2, R3, and so on) changes from beat to beat. It's this heart rate variability that gives us resilience in life. Source: Dreamstime © Z_i_b_i.

What's important to our discussion is that the distance from the top of one large spike (called an R wave) to the next is not always the same; it varies from beat to beat. While it may look like the space from one peak to the next is identical, when we measure the intervals we find that the distances between them change. And it's a good thing they do, because this is where our resilience in life begins.

The greater the variability between beats, the greater the resilience we have in facing life stresses and changes in our world.[21] Because we're measuring the variability between heartbeats, the measurement is called just what we'd expect: *heart rate variability* (HRV). HRV is measured in very small units of time called *milliseconds*, and the difference between one heartbeat and the next may be a matter of only a fraction of a millisecond.

As children we have a high HRV, and it makes perfect sense that we would. When we're young, and exploring and adapting to the world, our bodies need a way to adjust to what we find. And they need to adjust quickly. The first time our fingers discover what hot water from the kitchen faucet is all about, for example, or when we find out that not all dogs are as friendly as the one in our living room, we need to respond quickly. The heart's capacity to alter its rhythms—our HRV—and send blood to where it's most needed is how we're biologically wired for the fast response that's key to our survival.

It's the signal the heart sends to the brain that creates the coherence described previously. Just to be clear, the heart and brain are always in some state of coherence. In the chaos of daily life and the presence of negative emotions, our coherence levels may be low. Through simple exercises such as the one described earlier in this chapter, "Discovering and Resolving Your Unresolved Stress" (see page 183), and the one below, we can shift key parameters in our bodies to create higher levels of coherence. There's a direct connection between the HRV in our bodies, our level of coherence, and the resilience we have when facing the extreme changes in our world today, or the extreme loss that would come with multicentury life spans. The connection is this: The greater our level of coherence, the greater our HRV and our resilience will be.

Key 39: More heart-brain harmony (coherence) leads to greater life resilience.

Many of the recent discoveries regarding heart coherence, heart intelligence, and how to use both of them in our lives have been made by the scientists at the Institute of HeartMath. Through peer-reviewed research, IHM has shown beyond any doubt that two factors relate directly to personal resilience in our everyday lives:

- Our emotions can be regulated to create coherence in our bodies.
- We can use simple steps to create coherence on demand in our everyday lives.

Working with some of the most prestigious organizations and innovative researchers in the world, IHM has developed a simple system known as Attitude Breathing® that allows us to apply the discoveries they've made in their laboratories easily in our everyday lives. According to the researchers, the main benefit of this technique is that "the heart will automatically harmonize the energy between the heart, mind, and body, increasing coherence and clarity."[22] At IHM they've distilled the shift in emotions that creates the greatest levels of coherence into the following three simple steps, which are adapted from *Transforming Stress* by Doc Childre and Deborah Rozman.[23]

EXERCISE

Three Steps to Personal Resilience: Attitude Breathing®

Step 1. Recognize an unwanted attitude—a feeling or attitude that you want to change. This could be anxiety, sadness, despair, depression, self-judgment, guilt, anger, overwhelm—anything that's distressing.

Step 2. Identify and breathe in a replacement attitude. Select a positive attitude and then breathe the feeling of that new attitude in slowly and casually through your heart area. Do this for a while to anchor the new feeling.

Step 3. Tell yourself to take the "big deal" and drama out of the negative feeling or attitude. Tell yourself: *Take the significance out.* Repeat this over and over as you use Attitude Breathing® until you feel a shift or a change. Remember that even when a negative attitude feels justified, the buildup of emotional energy will block up your system. Have a genuine "I mean business" attitude and heart intent to really move those emotions into a more coherent state.

As you keep practicing, you will start to create new neural pathways and old trigger attitudes and resistances start to release.

The best science of the modern world has revealed that we do, in fact, begin healing the moment we are born. And our healing begins at the most fundamental level of our bodies, with the DNA itself. Now it's up to us to embrace our healing and the very real possibility of multicentury lifetimes, or even immortality, if we choose.

Regardless of how long we live in this world, however, the ability to self-heal also enables us to experience a quality of life that determines the success of our relationships, jobs, and careers. It's our ability to make these choices, in a way that no other form of life is capable of, that can make the difference between succumbing to the circumstances of fate and rising to our greatest destiny.

chapter seven

We're "Wired" for Destiny

From Evolution by Chance to Transformation by Choice

"Destiny is not a matter of chance; it is a matter of choice. It is not a thing to be waited for, it is a thing to be achieved."

— William Jennings Bryan (1860–1925), American politician

Sometimes the best way to understand a complex idea is through the eyes of someone who sees the world simply. The wisdom of Forrest Gump, the title character played by Tom Hanks in the 1994 film of the same name, is a perfect example of this kind of vision. When Gump is asked about the role of destiny in our lives, his timeless words ring just as true today as when he spoke them on the big screen for the first time, over two decades ago. "I don't know if we each have a destiny," he says, "or if we're all just floatin' around accidental-like on a breeze. But I think maybe it's both."[1]

Gump's philosophy precisely describes what personal transformation is all about. As individuals we each have a destiny that awaits us as the fulfillment of our greatest potential. Our destiny is ours, however, only if we act. Through the choices we make in each and every moment in our lives we claim this personal destiny. The certainty of knowing who we are, and how we fit into the world, is the compass that can guide us as we make our choices one day at a time.

TWO PATHS TO UTOPIA

A wave of bold novels written at the turn of the 20th century offered a glimpse of our collective future if the events of their day continued uninterrupted. Each book describes a time when humans have triumphed over the natural and technological problems that were common at the time of the writing. What sets the books apart is the way the problems were solved.

Arguably the best known of these books is Aldous Huxley's *Brave New World*, published in 1932.[2] Huxley's vision of the future is set in London in the year 2540, six centuries ahead of his own day. In that future, humankind has evolved beyond the limitations and suffering of the past. Huxley describes a world of peaceful co-existence where the population is limited to the number of people that Earth can comfortably sustain, where war is a thing of the past, where everyone is happy and has everything they need, where each person is educated, and where disease no longer exists and everyone remains perfectly healthy up until the last day of their lives. But the future that he describes comes at a very high price. To achieve the bliss of Huxley's utopia, the very qualities of human life that we most value and cherish dearly have become casualties of the solution.

The optimized population, for example, has been made possible because natural human reproduction has been abolished. In the brave new world, human embryos are created and incubated in controlled facilities. They are selectively engineered—genetically designed—to reach specific IQ levels that qualify them for specific jobs that are designed for them under the precision of a caste system. Everyone performs the work that matches the aptitude they were designed with, and they're happy to do so because that's all they know. They're educated only to the level they need to be to do their job. Everyone receives exactly the same pay, so there is no jealousy. People know from childhood when they will die, because their lifetime is programmed to end at 60 years of age. But there is no fear of dying, and no sadness when a friend or acquaintance dies, because the emotional bonding of partnerships and families that's the source of such pain has been abolished as well.

Quiet and contemplative moments are discouraged, and people are encouraged to spend their leisure time in groups, enjoying activities and eating foods that make them feel good. And while recreational sex is encouraged, sex for the purpose of love is obsolete. All of this unfolds under a global form of government that is led by 10 emotionally neutral leaders known as world controllers.

Huxley's point in writing the book is to show that while it's possible to solve the problems that have plagued humankind since the beginning of time, the catch is to do so without drowning the very spark of individuality, creativity, and self-expression that makes us who we are and gives meaning to our lives.

Huxley's book was inspired by earlier literary works exploring our future, such as H. G. Wells's *Men Like Gods,* published in 1923.[3] While *Men Like Gods* was written nine years before *Brave New World,* the story takes place in a world that's 3,000 years into the future. Through a fluke of nature, the main character of the story, a London-based journalist named Mr. Barnstaple is transported in his automobile to a future Earth in the year 4923, when there is no world government and religion and politics are only distant memories. They're all part of a mysterious past, known as "the time of confusion."

In Wells's future, the people of the world have embraced a form of education and government based upon five principles of liberty: 1) privacy, 2) free movement, 3) unlimited knowledge, 4) truthfulness, and 5) free discussion. Mr. Barnstaple finds that this new world is so attractive that, not surprisingly, he wants to stay there for the rest of his life.

The plot of the book takes a turn, however, as he learns that the best way he can assure the future he's discovered is to return to the familiar world that he came from and share what he's experienced. In doing so he plants the seeds and sets into motion the ideas that make such a future possible.

PARALLELS TO TODAY

I'm sharing details from both of these tales to contrast their visions of what's possible for our civilization in the future. Both authors concocted worlds where the big problems of our day have been solved. In both visions, war has become obsolete. Both books describe a time when people are happy and healthy and have transcended the extremes and perils that we face in our world today. The point is that each book describes a very different path to achieve these outcomes:

- One is at the expense of the values that give meaning to our lives and the expression of what it means to be human.
- The other is through the cultivation of the very freedoms that make our creative expression possible.

The parallels between these books and where we already find ourselves headed in the contemporary world are unmistakable. We're living in a time of extremes. We're faced with choices that are not unlike those described at the turn of the last century, choices regarding population; social, educational, and financial equality; and sustainable ways of living. *The difference is that we've only just arrived at the crossroads that will determine what we want our lives to look like and which type of future we will choose.*

Key 40: We still have the opportunity to create a healthy future by defining the values that we cherish *before* we implement solutions that cause irreversible harm to us and to our planet.

This is where the question of who we are comes in. Once we have answered this question, I believe the values that will lead us to fulfill our greatest destiny are a natural outcome. Mastering the remarkable potentials of our bodies can empower us in ways that are highly beneficial to us as individuals and collectively as a species, as well as to all life on earth. Expressing these potentials makes us resilient and gives us avenues through which to solve our most pressing challenges.

Now that science has unlocked the secrets to some of nature's most closely guarded truths—such as quantum reality, the genetic code, and splitting the atom—knowing the secrets of our own capabilities is critically important. For the first time in recorded human history, our access to these secrets gives us the power to chart our collective destiny, or to seal our collective fate and to do so in a single generation. This is exactly the situation that Aldous Huxley described in *Brave New World*.

It's precisely *because* we've unlocked so many of nature's secrets and become so powerful over life on earth that we now must figure out how those secrets fit into our lives and choose our course carefully. And while we may ask ourselves this question casually from a philosophical point of view, it's been the serious topic of passionate ethical debate in scientific circles for decades.

WHAT GIVES US THE RIGHT?

From the mid-1970s until the early 1990s, I was privileged to work among teams of brilliant earth scientists and aerospace engineers who were developing some of the most advanced technology the world has ever seen. For corporations and universities alike, this was a time of tremendous momentum, as America was redefining its dependence upon foreign oil as well as developing futuristic technologies during the ongoing Cold War and space program. Not surprisingly, this period of such intense research was accompanied by equally intense introspection. Scientists were exploring the limits of their newfound capabilities to alter life, the climate, and our planet at a level historically left to God and nature. The degree of responsibility that comes with such awesome power often sparked heated debates about our moral right to use such technologies—debates that I enthusiastically joined at every opportunity.

The discussions that erupted in front of office vending machines and laboratory drinking fountains, and that often continued into restrooms and cafeterias, generally followed one of two schools of thought. One school believed that our ability to "tweak" the forces of nature was, in and of itself, license to explore those technologies to their fullest. In other words, because *we can* modify weather patterns and create new forms of life, *we should* do so, just to see where the technology might lead us. A common justification for this thinking was, "If we weren't meant to do such things, we would never have discovered the secrets that made them possible."

The second school of thought was more conservative, suggesting that just because we have the ability to engineer life doesn't mean we have the right to do so. To supporters of this approach, the forces of nature represented sacred laws that should not be tampered with. To customize the genetic code of our children before they're born, for example, or to adjust global weather patterns to suit our needs is off limits, they argued. Tampering with nature, to them, would violate an ancient, basic, unspoken trust.

Although this "trust" is not necessarily spelled out word for word, this second school of thought would argue that if we cross the line from *user* to *creator,* we place ourselves in forbidden territory, possibly with unintended consequences. Some scientists actually draw upon Huxley's *Brave New World* to illustrate the slippery slope down which such a path could lead. The analogy often used was an automobile's speedometer. Just because the dial may indicate speeds up to 160 miles per hour, this does not necessarily imply that we should drive our vehicles that fast!

It's precisely the metaphor of a speedometer that I feel illustrates a third, as yet unidentified possibility. If a speedometer indicates that a vehicle is capable of traveling at 160 miles per hour, in all likelihood someone will attempt to drive at this speed at some point in time. After all, it's human nature to test limits, push boundaries, and carry our capabilities to the extreme. The key here is that when we do test our limits, we should have the wisdom to determine the time, place, and conditions of the test.

We can find a deserted stretch of road, with a good surface, on a day when the weather is good, and minimize the possibility of injuring ourselves and others—or we can act on an impulse and test the limits of a vehicle on a busy freeway, endangering ourselves and risking the lives of those around us. In either scenario, the boundaries are tested. One is done responsibly, the other carelessly.

The same principle of responsibility must apply to the way we push the limits to engineer the forces of creation. We live in a world where we've entrusted science and scientists to lead our journey of exploration, a journey from which there is little likelihood that we'll ever turn back. The choices that we are making right now, with their assistance, regarding fossil fuels, climate, health, healing, and the global economy, impact each and every one of us on a daily basis. They impact our 401(k)s, IRAs, and retirement plans, and factor into whether or not our children will be able to afford an education. They affect the kinds of industries that will thrive and what jobs will be created in our communities. They determine the future of our health-care system and whether our doctors will simply prescribe drugs when our poor lifestyle choices catch up with us or help us create lives and lifestyles where we need fewer drugs.

By honoring and basing our lives upon core values, we may ensure a future that leads us to our greatest destiny rather than to mutual destruction. As we shift from the old human story of separation, competition, and conflict to the new story of connection, cooperation, and sharing, we find ourselves at a precipice where we must choose the values that we cherish most, both as a species and as individuals in our everyday lives. We're at a rare "sweet spot" between old and new ways of thinking, where we can still choose the future we want and the path to get us there. And it all comes down to what we think when we answer the question *Who am I?*

THE GOOD NEWS IS THAT THERE'S A LOT OF GOOD NEWS!

There's a lot of good news in the world. Though it's often drowned out by the noise of the media machine that spins our attention to the crisis of the day, the good news exists nonetheless. Examples include the already-existing solutions for the personal and global issues that challenge our lives. So let's begin with the headline that should be appearing on the front page of every Sunday paper: The simple truth is that our biggest problems are already solved!

Key 41: We already have all of the solutions—all of the technological solutions—to the biggest problems that we face as individuals, communities, and nations.

Contrary to the thinking that we need to gather the scientists, engineers, spiritual teachers, and political leaders of the world together in one room somewhere to figure out how to create the best world and healthiest lives possible, the good news is that it's already happened. We have already created the think tanks, brain trusts, and policy centers needed to achieve precisely such goals; we began doing so over a century ago. And these institutions have found answers!

From the Carnegie Endowment for International Peace, which was founded in 1910 specifically to "hasten the abolition of international war, the foulest blot upon our civilization,"[4] to the Tellus Institute, founded in Boston, Massachusetts, in 1976 "to advance the transition to a sustainable, equitable, and humane global civilization," the framework is already in place to identify options for developing a globalized world. The current focus of the Tellus Institute's research, for example, is to use advanced scientific techniques to identify possible scenarios for humankind's future. This includes the search to identify a future that's sustainable and equitable and the policies, actions, and choices to get us there.

The point I want to make here is that we've already done the work. We've identified the big solutions and we already know what's possible when it comes to addressing problems such as food security, abundant energy, sustainable economies, and a health consciousness based upon self-healing. And it's a good thing we have these solutions now, because we certainly don't want to wait until the moment they're needed to begin the search to find them.

Let's take a high-level look at some of the solutions I'm talking about, so you have a sense of what I mean here.

We already have the food we need. We already have all the food we need to feed every mouth of every child, woman, and man on the face of the earth today. According to the United Nations World Food Programme, barring an extreme and unforeseen event such as an asteroid colliding with the Earth or a global nuclear war, "There is enough food in the world today for everyone to have the nourishment necessary for a healthy and productive life."[5] A lack of food is not the reason that there are approximately 925 million hungry people in the world, amounting to "more than the combined populations of the United States, Canada, and the European Union."[6]

What's lacking is the thinking and the leadership that makes it a priority for the food that we already have, to get to the places where it's needed most. To be clear, I'm not suggesting that this needs to be American leadership, or the top-down leadership of any single country, for that matter. What I am saying is that it's the acceptance of the status quo that makes the tragedy of starvation possible in a world where food is abundant and the technology to get that food where it's needed exists.

We already have the energy we need. We already have the technology to bring electricity into the home of every family on the face of the earth—clean, green, sustainable energy that emits *zero* greenhouse gasses. And we've had the technology for over 60 years.

When we talk about energy, the tendency is to base our discussions on our energy experience of the past, one largely based upon the burning of fossil fuels: first coal and then oil and natural gas. Realistically these forms of energy will probably remain in the world's energy equation for the foreseeable future. However, they don't need to. We already have solutions that make the energy sources of the past obsolete. And just as the world is changing faster than even the "experts" could imagine, the shift away from burning something like oil or coal to drive a turbine is approaching quickly.

Sources of energy fall into two main categories.

- *Conventional renewable energy.* At the mention of renewable energy, the "big three" sources generally come to mind—solar, wind, and hydro—and, to a lesser extent, geothermal. Rather than thinking of any one of these sources as *the* single solution to the world's energy needs, it makes sense to think of them locally, and consider what each local environment can offer and sustain. While centralized, powerful, and reliable energy sources may be good to run the hospitals, schools, high-rise office buildings, and apartments in some big cities, there are places where local sources can supply—and in some cases replace—large, centralized systems. America's desert Southwest is a perfect example of what I mean.

The Four Corners area of Arizona, Colorado, New Mexico, and Utah is well known for the long days of sunlight, and the quality of sunlight, that it receives nearly every day of the year. Albuquerque, the largest city in New Mexico, for example, experiences an average of 278 days per year of sunshine, and some of the smaller communities in the northern valleys of the state average 300 sunny days each year. In places like these, it makes perfect sense to use solar energy to supply homes, offices, and small businesses with the electricity they need during the daylight hours that they typically operate. In the same region, however, there are other supplemental forms of power generation that can also be tapped. In addition to the sunlight in Four Corners, its weather patterns provide conditions that make wind energy a viable alternative to fossil fuels, for example.

- *Unconventional but proven energy.* During the super-secret Manhattan Project of the mid-20th century, the race was on in the United States to find the mineral that could run the nation's nuclear reactors and produce plutonium by-products that could be made into weapons during the Cold War.[7] While most people are generally aware that this is the case, they're also surprised to learn that another mineral was discovered during this research that had many of the qualities of uranium as a fuel source but lacked the harmful side effects and dangerous by-products. The element is *thorium,* which is number 90 on the periodic table. Thorium was bypassed as a fuel source

> largely because it cannot be made into weapons, as uranium is today.
>
> A thorium generator works on a principle that is the opposite of a conventional nuclear reactor. *In a thorium generator, the warmer the liquid becomes, the slower the rate of nuclear reactions.*[8] This means the stuff that's producing the reaction *is the same stuff that prevents* any further reaction at high temperatures. This difference means that a Fukushima-like meltdown could never happen with a thorium generator. The physics make it impossible.
>
> Many people are surprised to learn that thorium energy has advanced beyond theory. It already exists.
>
> A number of thorium generators have already been built and are being used for research and commercial applications in countries that include India, Germany, China, and the United States. In the United States there have been two thorium generators: the Indian Point facility in New York State, which was operational between 1962 and 1980, and the Elk River facility in Minnesota, which was operational between 1963 and 1968.[9] While we need more research before thorium technology can meet the large-scale needs of the world, it holds the promise of a clean, abundant, and relatively safe alternative to tide us over while we search for the ultimate source of energy.

The next generation of power will be based upon infinite or "free" energy. The principles for such energy were discovered over a century ago and are the focus of people seeking next-generation alternatives to fossil fuels.

ECONOMIES BASED UPON SHARING RATHER THAN SCARCITY

The advent of modern technology is changing traditional thinking when it comes to the role of business and services in the modern world. The historic model has been that whatever product or resource is needed is owned by some party. That party then makes their goods and services available at a cost that covers their expenses and provides them with a profit. The need for rules and regulations in this model is obvious. The volume of regulations, and the opportunity to evade those regulations and "game the system," has made this kind of economy burdensome and ruthlessly competitive.

A new model is emerging that addresses at least some of these issues. It's based upon what's been called a *peer-to-peer* or a *sharing-based* economy. A sharing-based economy challenges the traditional ideas of ownership and relies on shared production by the same people that are using the service. In this way the need for hurtful competition and hoarding of what's valued no longer makes sense.

The noncorporate taxi services Uber and Lyft and the noncorporate hotel alternative Airbnb are examples of the new sharing economy. While the nuances of how these new models operate are still hotly debated, the bottom line is that they've emerged from the very people that are using them and have created a welcome source of income in difficult economic times. In 2013, for example, it is estimated over 3.5 billion dollars of revenue was generated by new businesses in the sharing economy.[10]

THE SILENT CRISIS

When we see examples of solutions, such as in the preceding paragraphs, and realize that we already have them, there's a single question that commonly comes to mind. I hear this question from live audiences throughout the world. The question is this:

Where are these solutions today? The answer often surprises my audiences. It's about a crisis that's seldom acknowledged, yet it creates the greatest hurdle that we face in our lives.

Our crisis is a silent one. It's rarely identified in mainstream media. There's no chapter in our college textbooks that describes its power and the huge role that it plays in our lives. Yet it remains an invisible wall between us and every one of the good news solutions that we could be benefiting from today.

Our silent crisis is a crisis of thinking. We have yet to shift the thinking in our lives to make room for the solutions in our world. When we consider this, it makes perfect sense. How can we embrace the new ideas and new solutions in our lives if we're clinging to the old ideas and solutions of the past? Put another way, how can we make room for the new world in our minds and in our hearts if we're filled up with the images, emotions, and expectations of the familiar world of the past?

Key 42: The greatest crisis we face as individuals and as a society is a crisis of thinking. How can we make room for the new world that's emerging if we are clinging to the old world of the past?

It's precisely for these reasons that the way we think of ourselves, *including our origins,* is now front and center in the decisions we're making when it comes to our everyday lives and our future. How do we implement the solutions that already exist, and do so in a way that honors the values that we cherish as individuals, families, societies, and nations? So far, the science of the modern world is leading us in the wrong direction.

DANGEROUS CONCLUSIONS

In October 1988, renowned astrophysicist Stephen Hawking summed up the traditional scientific view of how we fit into the big picture of the universe. In the German weekly publication *Der Spiegel,* he was quoted as saying, "We are just an advanced breed of monkeys on a minor planet of a very average star. But we can understand the universe. That makes us something very special."[11]

I remember my reaction when I first read these words from a man that I've always respected and held in high esteem. After all, Hawking was the man who wrote the 1988 blockbuster book *A Brief History of Time* that brought the complex ideas of cosmology and time travel into the living rooms of everyday families and made the idea of a black hole part of our vocabulary. While my sense is that Hawking wanted to convey that we are "special," he did so coming from the perspective of the science that tells us we're not. My reaction to his statement regarding us as "just an advanced breed of monkeys" was immediate. *Speak for yourself, Stephen Hawking!* I thought. *Maybe that's* your *story, but it certainly isn't mine!*

WHEN SCIENCE GETS IT WRONG

In my opinion, Hawking's statement regarding us as "an advanced breed of monkeys" is irresponsible. It's not based in fact. And I believe it's dangerous. It's a perfect example of how modern science has tried to remove the humanness from our human story. In sharing his words, Hawking is telling us something about himself personally, and revealing his own worldview. Either 1) he is uninformed and not aware of the latest fossil and genetic discoveries that make his statement false, or 2) he is informed and aware of them, and he's chosen to ignore the facts.

And if Hawking has chosen to ignore the facts, I can only speculate as to why he would make such a choice. Maybe it's to preserve the status quo when it comes to the story of human evolution. Or maybe it's something more personal. Maybe it's easier to make sense of the extremes in our world, and what happens in our lives, if we think of ourselves as "advanced monkeys." If we don't acknowledge the facts of our origins, the extraordinary abilities that are inherent in our existence, and that we are biologically wired to regulate our extraordinary abilities, then we remain the powerless victims of our biology. We are left to accept that whatever happens to us is somehow nature's will and beyond our control, rather than accepting responsibility for the world and our lives as we find them.

As extreme as Hawking may sound in his statement, however, he's not alone in his thinking. Other established scientists have taken a similar view when it comes to human evolution, and some in such a ferocious way that it leaves me wondering why they would continue to defend an obviously obsolete view so enthusiastically.

FALSE BELIEFS AND DANGEROUS CONCLUSIONS

Biologist and evolutionist Richard Dawkins is a very public example of what I mean here. Dawkins goes a step further than Hawking when he states, "It's absolutely safe to say that if you meet someone who claims not to believe in evolution, that person is ignorant, stupid, or insane."[12] While Dawkins isn't clear as to whether this statement is directed toward the theory of evolution in general or human evolution specifically, in either case these are dangerous words that represent a dangerous thinking—especially coming from a prominent scientist and university professor with such a visible presence on the world stage.

The reason Dawkins's words are so dangerous is that they chastise people for expressing curiosity and denounce the very essence of the act of scientific exploration. In his statement, Dawkins goes beyond professionally criticizing anyone who doesn't agree with him and evolution theory to publicly demeaning and even questioning the sanity of someone who feels the current scientific paradigm isn't convincing enough to buy into. I believe the thinking that Dawkins and others of his kind promote is dangerous for another reason as well, which has to do with how their reasoning leads us to think of other people and ourselves.

KILLING OUR UNIQUENESS

Among the extremes that we face in our lives today are the highly charged environments of human hatred. It's difficult to talk about. It's hard to believe how deeply into our lives it reaches. And yet it's there. Hatred is real. And it's a part of everyday life. Much of the hate in the world stems from the fears that we have of one another. Whether it's based in reality or our perception of reality, fear of the unfamiliar is at the foundation of the hate that we see in our schools, in our workplaces, and on the streets of even the most beautiful cities in the world.

In such a volatile environment, the very diversity that biologists tell us has always been our strength in the past—things like race, religion, and culture—has now been hijacked and cleverly packaged into talk show sound bites and shared YouTube video clips sold to the public as the wedge issues that separate and divide us. These divisions happen to different degrees and at different levels within different societies.

As a testament to the power of skillful marketing, the effort to polarize us through our differences has seen a surprising degree of success. A large portion of the general public has bought into them. For example, a recent survey conducted for *NBC News* and *The Wall Street Journal* showed a marked decline in the way both white and black people in America view race relations. The study found that, "according to the poll, 45 percent of whites and 58 percent of African-Americans now believe race relations are very or fairly bad, compared with 2009, when only 20 percent of whites and 30 percent of blacks held an unfavorable view."[13]

Clearly when it comes to factors such as religion and race, the meaning we give to those qualities is tearing us apart at the core of our families, workplaces, schools, and communities. And while this kind of division may be new to the millennial generation (young people born late in the 20th century), recent history shows that this is not the first time this kind of division has happened.

A NEW NAME FOR KILLING WHAT IS FEARED

Historians describe the 20th century as the single bloodiest century in all of recorded history.[14] In World War II alone, for instance, approximately *50 million* people died in combat and from war-related atrocities.[15] And the deaths due to human atrocities continued even after the war was over, through the end of the century. By 1999, *80 million* men, women, and children of all ages had been lost in the 20th century to violence based on ethnic, religious, and philosophical conflicts—five times as many as were lost due to all of the natural disasters and the AIDS epidemic *combined* during the same period.[16]

I'm sharing these horrendous statistics because they are part of a thinking that led to an increase of a new kind of atrocity in the last century. While acts of atrocity had certainly happened in the past, they reached such a magnitude in the 20th century that they had to be given an official name so they could be defined and made illegal.

In 1948, the United Nations adopted the term *genocide* to describe this kind of killing, as well as to make it possible to clearly define and outlaw mass murder in global policies. The act of genocide was defined as "an intent to destroy" societies or the populations of entire geographic regions based upon ideas of race, religious beliefs, or bloodlines.[17] The thinking that is used to justify genocide and make it possible is a poignant example of where false science can lead.

WE'VE SEEN THIS BEFORE

The thinking underlying contemporary genocides, and specifically spelled out in some, is directly linked to Darwin's false assumptions and how his ideas have been accepted, embraced, and perpetuated by modern science, even when they are proven false. California State University history professor Richard Weikart sums this up when he writes:

> Darwinism undermined traditional morality and the value of human life. Then, evolutionary progress became the new moral imperative. This aided the advance of eugenics [the belief that selective breeding, and the elimination of "misfits," can create an ideal human race], which was overtly founded on Darwinian principles. . . . Some prominent Darwinists argued that human racial competition and war is part of the Darwinian struggle for existence.[18]

This thinking is mirrored in the ideas of philosophical works such as the infamous Little Red Book, officially titled *Quotations from Chairman Mao Tse-Tung,*[19] and *Mein Kampf,* the book that detailed Adolf Hitler's worldview.[20] Both were used as justification for the brutal killings that took a combined toll of at least 40 million people in the genocides of the mid-20th century.

Sadly, divisive thinking hasn't disappeared with the passing of time. Since 1945, genocides have continued in places such Cambodia, Rwanda, Bosnia, and Sudan. These are well-documented tragedies that tell us the thinking that justifies mass killing is still present today.[21] And any doubt that we've somehow evolved beyond the thinking at the root of genocide quickly disappears with the well-documented tragedies of ISIS and the 21st-century genocides occurring in Africa and the Middle East.

In *Origin of Species,* Darwin clearly states his belief that the "weeding out" of the weakest members of the species he observed in nature applies to humans as well:

> It may not be a logical deduction but to my imagination it is far more satisfactory to look at such instincts as the young cuckoo ejecting its foster-brothers, [or] ants making slaves . . . as small consequences of one general law leading to the advancement of all organic beings—namely multiply, vary, let the strongest live and the weakest die.[22]

In *Mein Kampf,* Hitler clearly paraphrased this idea:

> In the struggle for daily bread all those who are weak and sickly or less determined succumb, while the struggle of the males for the female grants the right of opportunity to propagate only to the healthiest. And struggle is always a means for improving a species' health and power of resistance and, therefore, a cause of higher development.[23]

Later in life, Darwin had second thoughts with respect to some of his earlier statements regarding "survival of the strongest" in *Origin of Species.* Contrary to his early conclusions regarding superior individual strength, his later works described survival strategies in nature based on unity and cooperation, rather than natural selection and survival of the strongest. In his next major book, *The Descent of Man,* he summarized his observations: "Those communities which included the greatest number of the most sympathetic members would flourish best and rear the greatest number of offspring."[24]

Although Darwin may have seen the light in regard to his false assumptions of competition and struggle, it might have been too late. *Origin of Species* was already a classic text as the foundation of a mind-set that today is used to steer us away from reliance upon our natural instincts of cooperation and goodness.

NATURE'S RULE: COOPERATION

Early in the 20th century, Russian naturalist Peter Kropotkin reinforced Darwin's later work with his own observations. Just as Darwin had observed the effects of evolution firsthand among species of birds during his voyage of discovery in the 1830s, Kropotkin made his own observations during scientific expeditions to one of the harshest environments in the world: northern Siberia. *He described how he'd found that cooperation and unity, rather than survival of the strongest, are the keys to the success of a species.* In his classic book *Mutual Aid*, published in 1902, Kropotkin illustrated the benefits experienced in the insect kingdom through the instinctual ability of ants to live as cooperative rather than competitive societies:

> Their wonderful nests, their buildings, superior in relative size to those of man; their paved roads and overground vaulted galleries; their spacious halls and granaries; their corn-fields, harvesting and "malting" of grain; their rational methods of nursing their eggs and larvae, and of building special nests for rearing the aphides whom Linnaeus so picturesquely described as "the cows of the ants"; and, finally, their courage, pluck, and superior intelligence—all these are the natural outcome of the mutual aid which they practise at every stage of their busy and laborious lives.[25]

John Swomley, professor emeritus of social ethics at the St. Paul School of Theology in Kansas City, Missouri, leaves little doubt that it is to our advantage to find peaceful and cooperative ways to build the global societies of our future. Citing the evidence presented by Kropotkin and others, Swomley states that the case for cooperation rather than competition rests on more than just its benefit to a successful society. In a simple and straightforward fashion, he explains that cooperation is the "key factor in evolution and in survival."[26] In a paper published in February 2000, Swomley quotes Kropotkin, who states that competition within or between species "is always injurious to the species. Better conditions are created by the elimination of competition by means of mutual aid and mutual support."[27]

In the opening address at the 1993 Symposium on the Humanistic Aspects of Regional Development held in Birobidzhan, Russia, co-chair Ronald Logan offered a context for the participants to view nature as a model for successful societies. He directly quotes Kropotkin, who states:

> If we ask Nature: "Who are the fittest: those who are continually at war with each other, or those who support one another?" we at once see that those animals which acquire habits of mutual aid are undoubtedly the fittest. They have more chances to survive, and they attain, in their respective classes, the highest development of intelligence and bodily organization.[28]

At a later point in the same address, Logan cites the work of Alfie Kohn, author of *No Contest,* describing in no uncertain terms what his research had revealed regarding a beneficial amount of competition in groups. After reviewing more than 400 studies documenting cooperation and competition, Kohn reports his conclusion: "The ideal amount of competition . . . in any environment, the classroom, the workplace, the family, the playing field, is none. . . . [Competition] is always destructive."[29]

A growing body of ancient, scholarly, and scientific evidence suggests that in the absence of conditions that drive us to be animal-like in our actions (such as a *Mad Max* scenario, where there is a total breakdown of society, commerce, and medical care), when we're given the opportunity, we prefer to live peaceful and compassionate lives that honor the benevolent aspects of our species.

In other words, when the conditions we value in life are met—that is, when we feel safe, when we feel that our families are safe, and when we feel that our way of life is safe—we allow our truest nature to shine through in everything we do.

How can we know with certainty when these conditions have been met? Pulitzer Prize–winning poet Carl Sandburg answered this question in nine brief words: "Sometime they'll give a war and nobody will come."[30]

Key 43: A growing body of scientific evidence is leading to an inescapable conclusion: Violent competition and war directly contradict our deepest instincts for cooperation and nurturing.

As long as the diversity of our languages, religions, sexual orientations, and skin colors are falsely portrayed as flaws to be feared, people will turn against other people whose lives and beliefs differ from their own. They will shun, criticize, attack, and even try to destroy those whose ideals and beliefs they don't recognize in themselves. This is the common thread that connects each of the examples shared above. Each atrocity illustrates a lack of value for human life.

In a culture where life is valued and respected, none of the atrocities described here—or any of the countless additional ones that literally fill volumes in the Office of the High Commissioner for Human Rights at the United Nations—could ever happen.

KILLING WHAT'S DIFFERENT

The fact that race-based, gender-based, and religion-based atrocities pitting human against human have continued into the early years of the 21st century tells us that while we may have condemned the unthinkable acts of genocide we saw in the 20th century, we have yet to heal the thinking that makes these acts possible. Whether it happens at the level of nations, as genocide, or on a local level, as in the bullying in our schools or the resurgence of hate crimes in the United States in recent years, the fact that atrocities such as these exist at all is an indication that this way of thinking appears to be gaining momentum rather than becoming a thing of the past.

The following examples offer a brief glimpse of what I mean here. They represent only a sampling of an unsettling trend that's gaining strength in the world today.

Please note: This section was difficult for me, emotionally, to research and to write. My effort to reduce the countless numbers of victims under each category of hate crime to a single representative example in no way diminishes the suffering of the victims not mentioned or the pain that their families continue to experience. Due to the brutal nature of each example, I've chosen to summarize what transpired only from a high level of generality to 1) illustrate the thinking that underlies each example and 2) support my statement that this kind of thinking still exists today. Especially sensitive readers may want to skip ahead to the section titled "The Common Thread."

Cyber-based violence. While peer-to-peer harassment, shaming, and violence have probably existed as long as groups of young people have been confined together in classrooms of one kind or another, the degree of this kind of violence appears to be on the rise. There are different kinds of bullying that range from direct physical contact, such as hitting and spitting, to verbal attacks with no physical contact at all. Through the use of e-mail, Facebook, Twitter, and other online social networks, a new form of bullying appears to be on the rise: *cyberbullying*. Due to the growing use of social media among young people, cyberbullying is now documented to be very widespread.

According to the National Center for Education Statistics, since 2007 nearly one-third of all students ages 12 to 18 have been bullied while they were in school. A 2014 study conducted by the U.S. Department of Education reported, "During the 2009–10 school year, 23 percent of public schools reported that bullying occurred among students on a daily or weekly basis."[31] The statistics show that all forms of bullying, including cyberbullying, are dangerous. They all have painful consequences, some that can last well into adulthood, and some that are so painful that they lead students to take actions that are irreversible, such as suicide or murder.

On January 14, 2013, a distraught 15-year-old student named Jadin Bell walked onto the campus of an elementary school and hanged himself on the outdoor jungle gym. Jadin was a member of the high school cheerleading team and the victim of what was characterized as "intense" bullying through social media, largely because of his sexual orientation. His attempted suicide was unsuccessful at first, however, and he didn't die immediately. Instead, Jadin was found unconscious but alive and rushed to the nearest hospital, where he remained in a coma and on life support until he died on February 3, 21 days later.[32]

Jadin's suicide made national headlines and helped catapult the phenomenon of cyberbullying into the national conversation. His death powerfully illustrates how nonphysical bullying can have devastating emotional effects. According to Jadin's father, his son was "hurting so bad. Just from the bullying at school. Yeah, there were other issues, but ultimately it was all due to the bullying, for not being accepted for being gay."[33]

Sadly, Jadin's suicide is not an isolated incident. A growing number of young teenagers feel that taking their lives is the only way to cope with the humiliation that comes from cyberbullying. The nature of the bullying students endure ranges from insults about their appearance, weight, or physical features, and sharing of nude pictures that were originally taken in confidence to young girls being filmed during assaults and then being humiliated a second time as the videos are shared publicly on social media.[34]

Violence based on sexual orientation. Statistics compiled by the Federal Bureau of Investigation, the U.S. Census Bureau, the Pew Research Center, the Williams Institute, and demographic mapping website SocialExplorer.com were used to compare the number of hate crimes occurring in the United States targeting LGBT, Jewish, Muslim, black, Asian, and white people between 2005 and 2014. The result of the nine-year-long study was clear. As summarized in *The New York Times*, the study found that LGBT people "are twice as likely to be targeted as African-Americans, and the rate of hate crimes against them has surpassed that of crimes against Jews."[35] The savage killing of a young man in rural Wyoming provides a powerful example of the brutality that can stem from extreme thinking about sexual orientation, and that led to the later study.

Matthew Shepard was studying political science at the University of Wyoming in 1998. A gay man, he was enjoying an evening out at a local lounge on October 6 of that year with two other men who pretended to befriend him. At the end of the evening, they offered to give him a ride home and he accepted. But instead of being taken home, he was driven to a remote area where he was severely beaten, lost consciousness, and was left for dead. He was still alive, however, and in a coma, when a police officer discovered him at the remote location eighteen hours later. Doctors determined that the injuries to Matthew's brainstem were so severe that they could not safely operate on him. He remained on life support until he was pronounced dead on October 12, 1998.[36]

The high visibility of Matthew's story and of the trial of the men who were found guilty of his murder was due, in large part, to the antigay motivation of their actions.

Violence based on race. In June 1998, a man who had been hitchhiking one evening near his own town in rural Texas accepted a ride offered by three men, one of whom he was acquainted with. The hitchhiker, James Byrd, Jr., was black, and the men who picked him up in their car that evening were white. At least two of the three white men were self-described white supremacists. The events that followed and led to James's death were so brutal that they had to be censored by the national media in the public interest. It was this event, however, along with the hate killing of Matthew Shepard the same year, that led to the passage of a federal law named the Matthew Shepard and James Byrd, Jr., Hate Crimes Prevention Act, which expanded the 1969 United States federal hate crimes law to include crimes motivated by a victim's actual or perceived gender, sexual orientation, gender identity, or disability. The act was passed by the U.S. Congress on October 22, 2009, and signed into federal law by President Obama on October 28 the same year.[37]

Violence based on religion. In testimony presented at the British House of Commons in 2016, one of the British ministers read verbatim portions of an interview with Ekhlas, a 15-year-old girl living in northern Iraq, who, along with her family, followed the ancient Yazidi religion. Ekhlas's village was overrun by ISIS soldiers, she was captured and enslaved until she escaped from her captors. [38] She described how these men had come to her family's home, killed her father and two brothers, and then brutalized her and every girl in her village over the age of nine. The reason for their ordeal, she said, was their religion. "We were targeted because our religion and belief is different from theirs, and our humanity is different from theirs, because we believe in the angel Taus."[39]

Religion-based hate crimes are not limited to the Middle East. They are making a resurgence in other parts of the world as well, including Europe and the United States. Since 1996 the FBI has logged statistics for violence against people in the United States based upon their religious beliefs. The 2014 Hate Crime Statistics report states that 5,479 incidents of hate crime in general were reported in 2014. Of this number the percentage that were religion-based crimes against individuals was 17.1 percent.[40]

Interestingly this is very close to the percentage for crimes based upon sexual orientation (18.7 percent).

The study also shows that of the offenses reported, "Approximately 58.2 percent were anti-Jewish, 16.3 percent were anti-Islamic, and 6.1 percent were anti-Catholic."[41]

THE COMMON THREAD

There is a common thread that weaves its way through the examples of hate crimes I've just described. In following that thread, we're given insight into the kind of thinking that's tearing at the very fabric of our families, communities, and societies. In each instance, the brutality of the hate crime could happen only in the presence of a belief that the life of the victim had no worth.

Key 44: The brutality of hate-based crimes is possible only in a society where the value of human life has been lost.

Hate crimes are about something that goes much deeper than the taking of another person's life. They are rage-filled demonstrations of overkill—murders based upon an almost primal fear of the unfamiliar, combined with a belief that human life is common and expendable. And while the previous examples are extremes of where such thinking can lead when it's expressed outwardly toward other people, hate can be directed inward as well, demonstrating an extreme of a different kind.

Inward-directed abuse is sweeping through our schools and touching the lives of our sons and daughters, siblings, friends, mothers, and fathers. And it's affecting our young people in epidemic proportions. While it's happening in a way that's more subtle than the violent hate crimes I've described, the outcome is the same. The self-inflicted abuse of prescription drugs and alcohol often leads to the devastating loss of the people that we love the most.

The pain of losing a loved one to their inwardly expressed hatred is almost beyond expression. That pain is especially acute when a surviving family member struggles with unanswered questions and the feeling that if only they'd done something differently, their loved one would still be alive. Tara Lawley-Bergey, the older sister of Derik Lawley, describes this pain in an essay she wrote after Derik's death due to a lethal dose of the drug fentanyl to satisfy his heroin addiction.

TARA'S STORY

In an essay that was published by the Philadelphia affiliate of NBC in February of 2016, Tara describes how her brother had suffered from a heroin addiction for two and a half years.[42] She reveals that she never really knew why Derik began to experiment with heroin. But she speculates about what may have happened. Tara says her brother loved life. He loved those around him, especially his three-year-old daughter. But he didn't love himself. "Heroin helped Derik escape his reality; it put him into a daze that allowed him to forget," she writes.[43] And although he attempted to wean himself from his addiction at least five times, his efforts were unsuccessful.

Derik's body was found abandoned in a wooded alley a full day after he had been deceptively given fentanyl, a narcotic associated with anesthesia, when he believed he was receiving a familiar dose of heroin to satisfy his addiction. He died from the effects of the drug, which put him into such a deep sleep that it suppressed his breathing. Tara's pain in reflecting on her brother's experience is best described in her own words:

> My heart died the moment Derik took his last breath. His body lies in ashes as mine dies slowly from within. The darkness lingers and the nightmares loom into the light. The pain of losing Derik is unbearable, and I am living in the ninth circle of hell, my treachery being called an addict's sister. Siblings love each other regardless of their paths; they guide each other when they have fallen and are each other's shoulder to lean on. But I distanced myself from Derik's addiction; it made him a wicked man. I should have been there for Derik, to wipe the sweat of addiction off his brow when the wickedness came upon him time and time again. Or in the least, I should have called, wrote, or sent Derik love in a care package. But I ignored him, gave him the cold shoulder, and did not see the real person within his eyes. I practiced tough love when I should have just shown him compassion. That is my burden, my guilt, my pain to bear all the days of my life.[44]

The tragic story of Derik is a powerful testimony to a preventable death. It's also a story that unfortunately is not a rare one. Time and again, different parents from different communities, different races, and different religions ask the same question through their tears as they bury their sons and daughters. The question is why. "Why did this happen to *my* child?" And as different as their families are from one another, the answer to their question is the same. A man, woman, or teenager who values himself and finds a sense of worth in his life would never pump heroin into his veins, snort cocaine into the delicate tissues that breathe life into his body, or flood her liver and kidneys with so much alcohol that she loses consciousness.

Key 45: The destruction of an individual through the abuse of drugs and alcohol is only possible when the personal sense of worth is lost.

WE ONLY DESTROY WHAT WE DON'T VALUE

Environmentalist and author Rachel Carson summed up the thinking that leads to such heartbreaking and devastating experiences for families throughout the world when she said that we destroy what we don't value and that we can't value what we don't know.[45] Carson's observation beautifully describes the theme of this book and the crux of what we're up against today. And while experts attribute the uptick in person-to-person violence to everything from inequality between those who "have" and those who "have not" to religious intolerance among Christians, Jews, and Muslims, the real reason that lies at the core of all reasons for the increasing violence toward one another is the source of a difficult truth.

While we've created a wondrous society and a culture of advanced technology, we've done so at a tremendous cost. Somewhere along the way we lost the value that we place on a human life. And without that sense of value, life seems expendable. The treatment of textile laborers at the turn of the 20th century is a perfect example. Only days after dozens of textile laborers died in New York City's Triangle Shirtwaist Factory fire in 1911, worker and union activist Rose Schneiderman gave a speech describing how human life is undervalued:

> This is not the first time girls have been burned alive in the city. Every week I must learn of the untimely death of one of my sister workers. Every year thousands of us are maimed. The life of men and women is so cheap and property is so sacred. There are so many of us for one job it matters little if 146 of us are burned to death.[46]

Although Schneiderman made this speech over a century ago, the conditions that she described and the thinking that makes those conditions possible haven't changed all that much. We need look no farther than everyday headlines throughout the world to see how deeply the sense that life is "cheap" continues to play out in our daily existence.

- Between 2001 and 2012, the number of women in America killed by their former or current partners was 11,766, over twice the total number of American troops killed in the Afghanistan and Iraq wars combined, during the same period of time.[47]
- In 2013, disregard for safe conditions in a garment factory in Dhaka, Bangladesh, led to the collapse of the building and the death of over 1,000 people, making it the worst disaster of its kind in history.[48]

Key 46: Rachel Carson reminds us that we only destroy what we don't value and we can't value what we don't know. A lasting solution to the issues that divide us and the growing levels of bullying, hate crimes, and wartime atrocities is to instill in the new generation, and embrace within ourselves, the need to respect and value all life.

THE POWER OF SELF-WORTH

In our hypercharged environment of extremes, what we believe about who we are and where we come from holds a special place of sacred power. It's precisely these beliefs that hold the power to fragment our communities and polarize our nations so we engage in endless wars. Such beliefs also have the power to unify us. The deepest truth of our origins could give us a reverent sense of value for each and every human life.

This is why it's so dangerous to believe false science and lie to ourselves about where we come from. If it were true that we're "just an advanced breed of monkeys" and "ignorant, stupid, or insane" for believing anything other than the accepted doctrine of human evolution, then it would make perfect sense to live our lives in a way that reflects such a belief. In that world the pursuit of material wealth, the diversions of the mind, and the pleasuring of the senses become the highest priorities of life. In such a world, it would make sense to do whatever it takes to satisfy ourselves at any cost, in any way we can. Why not? After all, if we're simply the lucky outcome of nature's lottery of random mutations, why wouldn't we? Why wouldn't we swallow whatever chemical or tonic is available to numb us to the hurts of life? Why wouldn't we infuse our bodies with whatever manufactured drug or brain-altering substance is available to escape the insanity of war, the injustice of poverty, and the horrors of physical and emotional abuse? And why wouldn't we destroy anything or anyone that stands between us and getting what we need to live such a life?

Here's what I'm getting at: As long as we're led to believe that we're little more than an accident of nature, it will be easy to feel that there's nothing exceptional about us or our lives. From this sterile-sounding perspective, our story is simple, direct, and void of any deep meaning. We're born. We live. And we die. We're blips of life on nature's radar screen, just as billions of creatures have been before us.

The irresponsible words of renowned, high-profile scientists and public figures only make things worse for us by throwing fuel onto the fire of our differences and sense of insignificance.

FROM BAND-AIDS TO DESTINY

The potential to shift from merely identifying and condemning the atrocities that stem from a lack of self-worth and intolerance of our differences to embracing a destiny where such atrocities are only a memory of the past becomes possible when we consider the positive impact of our answer to *Who are we?* This answer, based upon the facts we now know to be true about ourselves, and particularly about the specialness of our existence, is the key to our new human story of life with a purpose.

In a culture where we embraced life's specialness, people wouldn't criticize, hurt, and kill one another or themselves with the ease and frequency that we see today. To do so would make no sense in light of what we know about our origins and what it means in our lives.

By embracing our specialness and the value of life at the foundational level of our selves and families, and by basing the education that we give our children on these uniquely human values, we could create the foundational shift—a complete sea change for people everywhere around the world—that leads to our greatest destiny of realized potential as a species. To do anything less is the equivalent of merely placing a Band-Aid on the open wound that's destroying our families, communities, and societies. In a culture where these values were embraced, Derik Lawley would have never succumbed to the temptation of taking heroin that cost him his life, James Byrd, Jr., and Matthew Shepard would be alive today, and the genocides of the 20th and early 21st centuries would never have happened.

On an individual level, in a culture that truly values life, this means that:

- A man who embraces the specialness of another life would never unleash his rage on a woman carrying his unborn child, on his existing children, or on anyone else he loves.
- The delicate balance that gives us our specialness would be honored. Men, women, and children would never poison their bodies with the alcohol and drugs that destroy the fragile systems that make their lives possible.
- Teenagers would never pull the trigger of a gun on a friend or themselves because life has placed them in what feels like an overwhelming situation.
- Someone driving a car would never pull out a gun and turn it on another driver for suddenly changing lanes in front of them to make a freeway exit.

On a larger scale, this means that:

- A soldier or rebel fighter who values the specialness of life would never brutalize another man or the man's wife and children simply because they don't share the same religious beliefs as him.
- A nation that embraces, shares, and teaches its children to respect and value life would never march onto the land of another nation to destroy its people's sources of water, food, and electricity or its schools and hospitals.

The way we think of ourselves and one another is at the very core of the greatest fears and the greatest suffering that we experience in our lives today.

While we can pass legislation to punish, send armies to coerce, and denounce human atrocities once they've happened, these are temporary fixes for conditions that can only change with a foundational shift in thinking—specifically in the way we think about ourselves, our origins, and the value we place on life on earth. And this foundational shift is what's lacking in the education that we offer our young people today.

Albert Schweitzer, winner of the Nobel Peace Prize in 1952, taught how vital it is for us to embrace a reverence for all life. "Only by means of reverence for life," he says, "can we establish a spiritual and humane relationship with people and all living creatures within our reach."[49] The reverence that Schweitzer talks about here goes beyond simply respecting life and includes our ability—*our duty*—to protect and defend all forms of life in need. "Only in this fashion [of reverence] can we avoid harming others," Schweitzer says, "and, within the limits of our capacity, go to their aid whenever they need us."[50]

We have an opportunity at this very moment in history—the "sweet spot" of choice described previously in this chapter—when we're determining the balance between what science and technology have made possible and how we implement those possibilities in our lives. It's the difference between Aldous Huxley's future, in which human creativity, individual expression, reproduction, and life itself are compromised for the sake of creating a homogenous and peaceful world, and the future described by H. G. Wells, in which humankind achieves a harmonious way of living and does so as the result of honoring and cultivating the values that we cherish. Whether we're talking about personal decisions related to health care, jobs, relationships, and careers, or more global issues, such as the need to find new sources of clean and sustainable energy and address the realities of poverty, social change, and the growing numbers of refugees created by oppression and war in the world, as complex as such issues appear at first blush, they all come down to the way we think about ourselves. For each of these issues, and a host of others, we're being challenged to identify the values that we cherish as humans and to claim those values as the guiding principle of our decisions. Once we recognize this, it's clear that only by embracing the worth of every person, and the value of each and every life, can we choose the destiny that holds our greatest potential.

Anglican bishop Desmond Tutu summed up this idea perfectly in a reminder that it's by sharing that which makes us unique—our capacity for love and compassion—that we discover our value. "Your ordinary acts of love and hope," he says, "point to the extraordinary promise that every human life is of inestimable value."[51]

So where do we start when it comes to creating a world that cherishes human life? Where do we even begin?

The first step is to embrace what we've discovered as the new human story.

chapter eight

Where Do We Go from Here?

Living the New Human Story

"One's destination is never a place but rather a new way of looking at things."

— Henry Valentine Miller (1891–1980), American Writer

The traditional answer to *Who are we?* is crumbling. It has to. The reason is that it's based upon information we now know isn't true. The key discoveries that are overturning the way we have thought of ourselves for 150 years are only the beginning of identifying a new human story. Once you've seen the discoveries, you cannot "un-see" them. You know they exist. They're already part of you. So you have to ask, "What now? How does this information fit into my life and what I want for myself, my family and friends, and the earth?" Finding answers to these questions begins with how closely you embrace what you've discovered.

Ultimately, what you do next comes down to a choice. It's your choice. What do you accept and what does it mean in your life?

When I'm faced with new and life-changing information, such as I have been when confronted by the evidence for a different scientific explanation of human origins than Darwin's original theory of evolution, there are three simple questions I ask myself to guide my choices.

Guidelines That Can Help You Make a Choice

1. Do I recognize that I have a choice?
2. Do I have the courage to choose?
3. Do I have the strength to follow through on the choice I've made?

When it comes to the very personal question *Who am I?* here's how the guidelines could be applied:

1. **Do I recognize that I have a choice** between believing the old story of human evolution and the new evidence that tells us evolution isn't our story?
2. **Do I have the courage to choose** to believe what the new science is telling us and to accept and embrace the new discoveries?
3. **Do I have the strength to follow through** and live what such a choice means when it comes to what I teach my children and the way I treat other people?

In any situation, your answer to these three simple questions can transform the way you think of your life and see yourself, and perhaps more importantly, it can change your actions. By developing the discipline to ask these questions before taking action, you automatically expand the number of options that are available to you. From the choices you make regarding diet and nutrition, honesty in relationships, and selecting health-care options to simply being more open to new possibilities for jobs, careers, and personal creativity, these simple guidelines will help you make your choices in a conscious and mindful way. You may be surprised to discover just how powerful you are when it comes to shaping what your life looks like today and creating a fulfilling tomorrow.

My reason for writing this book has been to share the new discoveries that give new meaning to the way we think of ourselves, and one another. But my reason for sharing this information goes beyond just wanting you to know the facts. The scientific evidence of our intentional origin through an as-yet-unknown intelligent external force gives new meaning to our existence. It carries us *beyond* survival of the strongest, struggle, and competition. It opens the door to a possibility that we may be related to something much greater than we've been led to believe in the past—and that we may have a cosmic history, a cosmic family, and a cosmic origin.

As a scientist, this idea sounds like the plot of a great science-fiction thriller at first blush. But it's where that thriller can lead that excites me. It's the possibility where we transform our lives and our world in the best way possible, and do so while honoring the human values we cherish most. In some respects this is the outcome described in H. G. Wells's book *Men Like Gods*—except that if we succeed, it will be happening 3,000 years earlier.

RETHINKING YOUR BASELINE BELIEFS

Now that you've had the benefit of reading this book and the discoveries it has revealed, I invite you to round out your reading experience by revisiting the questions I asked you at the beginning of Part I.

Before Chapter 1, I invited you to create a baseline for what you thought about evolution and its significance for your life and of what you believed about yourself. Now is a good time to reexamine those thoughts and beliefs to clarify if—and how—they have changed.

Opening the door to our greatest potential as human beings must begin with our willingness to embrace the fact that the potential for extraordinary things exists. After you've answered the following questions, I invite you to compare your answers to those you recorded at the beginning of this book. My overarching question to you is: Has what you've discovered transformed the way you think of yourself, your limits, and most importantly, your potential?

EXERCISE

Reassessing Your Baseline Beliefs

The Technique. Using single words or brief phrases, please write down your answers to the following questions as honestly as you can. For yes-or-no questions, circle your answer.

- **Questions about Your Origins.**
  1. Do you believe that the origin of life in general is the result of a chance event that happened long ago, as conventional science suggests?
     Yes No
  2. Do you believe that human life—*your life*—is the result of a chance event that happened long ago, as evolutionary theory suggests?
     Yes *No*
- **Questions about Your Potential.**
  3. Do you believe that you're designed to consciously influence the events of your life, the quality of your life, and how long you live?
     Yes No
     If you answered no to the previous question, continue to "Defining Your Beliefs" below.
     If you answered yes to the previous question, please continue here:
  4. Do you trust your ability to trigger self-healing in your body on demand when you need it?
     Yes No

5. Do you trust your ability to trigger your deepest states of intuition on demand when you need them?
   Yes No

6. Do you trust your ability to self-regulate your immune system, your longevity hormones, and your overall health?
   Yes No

- **Defining Your Beliefs.**

7. When I notice something unusual happening with my body (sudden aches or pains, an unexplained rash, a rapid heartbeat for no apparent reason, and so on) I now find myself *feeling* ____________________________.

8. When I notice something out of the ordinary happening with my body, *the first thing I will do is* __________________________.

The way you've answered each of these questions will guide you to understand how you currently think of your potential. These answers can also serve as a compass that indicates in what direction you may want to explore your personal growth. The key here is that your body can respond only to the fuel of the beliefs you embrace.

For example:

- If you believe that life in general, as well as your life in particular, is the result of a chance event that happened a long time ago, then this perception may be reflected in the choices you make in other areas of your life. For example, it's easier to discount the sacredness of life and the value of our experiences when we tell ourselves that we're the result of a lucky accident of biology that just "happened" to have occurred long ago. To embrace the growing body of evidence suggesting that we're the result of an intentional act—*when we really get that we're here on purpose*—we're left with a sense of awe and a deep appreciation for all life, everywhere. That appreciation is reflected in the way we think of ourselves, as well as the way we treat our friends, family, and loved ones.

- If you don't trust your body's ability to maintain your health, heal, and strengthen your immunity, or your capacity for intuition, then this perception may show up in how you respond to changes in your body. Do you become fearful at the first sign of something new or different in your body? When do you choose to see a doctor to interpret the signs that your body is showing you?

To be absolutely clear, there are no right or wrong answers to any of these questions. Your answers may be deep and personal reflections of the way you were conditioned to think of yourself. If that thinking has served you in the past and it continues to work for you today, then you are now consciously aware of the beliefs that guide you. But if you now find that you'd like to expand your relationship with your body, then your growth must begin with the beliefs that are the foundation of that relationship.

Perhaps it's no surprise that the more we know about ourselves—and the deeper we expand our awareness of our bodies' potential—the more purpose we sense to our lives. And I believe that's ultimately the goal for each of us: to discover and embrace our purpose while we're experiencing life's possibilities.

LIFE WITH A PURPOSE

Almost universally, the indigenous traditions of the world remind us that we're the products of a conscious and intentional act of creation and somehow part of a cosmic family, and that as we grow and mature in our understanding, our true heritage will come to have greater meaning in our daily lives. In ancient writings, varying from Sumerian cuneiform and Egyptian hieroglyphics to the carvings and pictograms discovered in the Mayan jungles of Central America, and in the spoken wisdom of Native North Americans and South Americans, our ancestors have told us that we're part of something vast and beautiful. As described in the scriptures of the world's most ancient traditions, we're given extraordinary abilities—godlike traits—that set us apart from all other forms of life and empower us to have connected, vital, and meaningful lives. We're reminded, as well, that we are stewards in this world, here to protect all life, and not masters born to dominate life.

It's through our extraordinary powers of intuition, empathy, and compassion that we are granted the privilege to be the earth's caretakers—an ability given to no other form of life. One of the greatest visionaries in history, Chief Seattle, a leader of the Suquamish people of the American Pacific Northwest, reminds us of our role in clear, eloquent, and direct terms. While the source of the following statement, often attributed to Chief Seattle, remains unconfirmed, the sentiment that it carries is timeless:

> Humankind has not woven the web of life. We are but one thread within it. Whatever we do to the web, we do to ourselves. All things are bound together. All things connect.[1]

The best science of the modern world seems to support the essence of this wisdom. Our extended neural networks and our ability to use our hearts, brains, and nervous systems to enhance our lives on demand are now scientifically documented. While there are scientists who may not share the self-empowering interpretations of the scientific findings that are offered in this book, what we can say with certainty is that there is nothing in the new discoveries that can deny that these capabilities exist in us or that would prevent these capabilities from being the result of an intentional design in the human genome.

While we may not fully understand where our advanced abilities come from, the evidence shows that our extraordinary intellect and our capacity for compassion, empathy, and deep intuition are no accident. They've been with us, as original "equipment," from our beginning. They are inherent in our nature and appear to have a purpose—they are a vital part of an intentional design.

THE REAL WORK

The world is on the move and our lives are moving with it. Now that you know what's in these pages, you cannot unknow what you've read. You can't simply close the book and forget about these discoveries regarding your origin or the immense power that lies within you. Although you've reached the end of this book, you've also reached the beginning of what comes next. This is where the real work begins. When you close this book, you'll be faced with a choice to discount what you've discovered about yourself or to embrace it.

Either choice takes effort. Either choice takes real work.

In his timeless book *The Prophet*, philosopher Kahlil Gibran makes a statement about the meaning of "work" that I remember reading when I was 10 years old. As a young boy living in a fatherless home with my single mother and younger brother in low-income, government-subsidized housing, Gibran's words provided a way of thinking that guided me then and has stayed with me as the cornerstone of my life's philosophy ever since. Gibran reminds us that work "is our love made visible."[2] To me, this has always meant that the effort that goes into any task is about more than the task itself.

If I've agreed to do something, then it's the meaning I give to whatever that something is that matters to me. My "love made visible" means I am 100 percent present and I give 100 percent of myself to whatever it is that I've said yes to doing. In other words it's not about what we do, it's about the way we do it. It takes work to be fully present, and from Gibran's point of view, that work is an expression of our love for the world, ourselves, and our families.

I'm a realist when it comes to the work it will take to embrace our new human story. It will take work to change the textbooks, computer files, teachers' class notes, and scientific displays in museums throughout the world. It will take work to teach the new human story to our children and then to theirs. It's through our work, our love made visible, that we chart our greatest human potential: the choice to move from evolution by chance to transformation by choice. The question now is, do we believe we're worth it?

Do we believe we're worth the work it will take to embrace the extraordinary potential that lies waiting within each and every human being? We won't need to wait long to know how we've answered. We'll know by the world we choose to leave for our children.

THE NEW HUMAN STORY IN 46 KEYS

Throughout this book, I have introduced the discoveries and facts that give us a reason to think differently about ourselves. To emphasize what I felt were landmarks within the book, I highlighted significant ideas and discoveries. But what may not have been obvious is that while each of these keys by itself summarizes an important theme, when the keys are read together, one after another, they tell a story. This story is the essence of the new human story. For your convenience, you can read the keys one after another below.

Key 1: In the presence of the greatest technological advancements of the modern world, science still cannot answer the most fundamental question of our existence: *Who are we?*

Key 2: Everything from our self-esteem to our self-worth, our sense of confidence, our well-being, and our sense of safety, as well as the way we see the world and other people, stems from our answer to the question *Who are we?*

Key 3: By allowing new discoveries to lead to the new stories they tell, rather than forcing them into a predetermined framework of ideas, we may, at last, answer the most important questions of our existence.

Key 4: New DNA evidence suggests that we're the result of an intentional act of creation that has imbued us with extraordinary abilities of intuition, compassion, empathy, love, and self-healing.

Key 5: The stories that we tell ourselves about ourselves—and believe—define our lives.

Key 6: When we change the story, we change our lives.

Key 7: For the first time in recorded human history, Charles Darwin's theory of evolution, published in 1859, allowed science to answer the big questions of life and our origin without the need for religion.

Key 8: While the connections between ancient primates and modern humans on the evolutionary family tree are believed to exist, they have never been proven as fact—they are inferred and speculative connections only, at this point in time.

Key 9: The discovery of an extraordinarily well-preserved female Neanderthal infant—dating back 30,000 years—and the comparison of her mitochondrial DNA to ours, tells us definitively that the earliest modern humans *were not* the descendants of ancient Neanderthals.

Key 10: Human chromosome 2, the second-largest chromosome in the human body, is the result of an ancient DNA fusion that cannot be explained by the theory of evolution as we understand it today.

Key 11: The 20 proteins that make the clotting of blood possible and the 40-plus components of the cilia (wiggly tails) that allow cells to move through a fluid are just two examples of functions that could not develop gradually over a long period of time as evolution suggests. In both examples, if even one protein or component part is missing, the function of the cells is lost.

Key 12: Humans appeared on earth with the same advanced brains and nervous systems we have today, and with the ability to self-regulate vital functions already developed, contradicting the corollary to evolution theory that nature doesn't "over-endow" with such features until they are needed.

Key 13: A growing body of physical and DNA evidence suggests that our species may have appeared 200,000 years ago with no evolutionary path leading to our appearance.

Key 14: An honest scientist, who is not bound by the constraints of academia, politics, or religion, can no longer discount the new evidence about our human origins and still remain credible.

Key 15: As part of our advanced nervous system, the heart partners with the brain as a master organ to inform the brain of what the body needs in any given moment.

Key 16: Ancient traditions have always held that the heart, rather than the brain, is the center of deep wisdom, emotion, and memory, as well as serving as a portal to other realms of existence.

Key 17: The discovery of 40,000 sensory neurites in the human heart opens the door to vast new possibilities that parallel those that have been accurately described in the scriptures of some of our most ancient and cherished spiritual traditions.

Key 18: Scientific documentation of memories carried from a donor into the body of a recipient through the heart itself—*memory transference*—demonstrates just how real the heart's memory is.

Key 19: The heart is the key to awakening deep intuition, subtle memories, and extraordinary abilities thought to be rare in the past, and to embracing these attitudes as a normal part of everyday life.

Key 20: Willingness to embrace a scientific assumption as fact, in the absence of evidence to support it, can lead us, and has led us in the past, to wrong conclusions when it comes to the way we think of ourselves and our relationship to the world.

Key 21: Renowned scientists tell us that it is mathematically impossible for the genetic code of life to have emerged through the process of evolution alone.

Key 22: Almost universally, ancient and indigenous traditions attribute our origins to the result of a conscious and intentional act.

Key 23: A growing body of evidence suggests that we exist as part of a living and vibrant universe rather than one simply made of inert dust, gas, and empty space.

Key 24: If we're the result of something more than pure chance, then it makes sense that our lives are about more than purely surviving. It implies that our lives have purpose.

Key 25: Our capacity for deep intuition, sympathy, empathy, compassion, and the self-healing that allows us to live long enough to share these capacities, are the needle of a compass that points us directly to our life purpose.

Key 26: *Intuition* is a real-time assessment that draws upon personal and past experience, sensory cues, and "street smarts," while *instinct* is a response that is "hardwired" into our subconscious as a survival mechanism.

Key 27: The emotional bond that exists between a mother and her children is now scientifically documented through studies that offer insights into the intuitive connection that we all can develop in our relationships.

Key 28: Intentional heart focus empowers us to consistently experience deep states of intuition when we choose, on demand.

Key 29: We can access our heart's wisdom through a process that can be summarized in five simple steps: focus, breathe, feel, ask, and listen.

Key 30: Intuition, sympathy, and empathy are the stepping-stones to compassion.

Key 31: Compassion is both a force of nature and an emotional experience that connects us with nature and all life.

Key 32: Telomeres are specialized sequences of DNA located at the ends of a chromosome that serve as a buffer to protect the chromosome's genetic information when a cell divides. With each cell division, the telomeres become shorter, until they can no longer protect the vital information of the cell, at which point the cell experiences old age, senescence, and eventually death.

Key 33: The purpose of the telomerase enzyme in our cells is to repair, rejuvenate, and lengthen the telomeres that determine how long our cells live.

Key 34: Our choices of lifestyle, including specific forms of exercise, specific dietary supplements, and reducing stress in the body, are key strategies documented to successfully slow and even reverse telomere damage and cellular aging.

Key 35: It's the *unresolved* stress in our lives that erodes our telomeres and steals from us the very thing we cherish most: life itself.

Key 36: Through our heart's wisdom we can ask for and receive insights into healthy alternatives to the unhealthy diversions in our life.

Key 37: In each moment of every day, we make the choices that affirm—or deny—life in our bodies.

Key 38: Heart-brain resilience is the key to emotional healing from the loss of family and loved ones that comes with extended life spans.

Key 39: More heart-brain harmony (coherence) leads to greater life resilience.

Key 40: We still have the opportunity to create a healthy future by defining the values that we cherish *before* we implement solutions that cause irreversible harm to us and to our planet.

Key 41: We already have all of the solutions—all of the technological solutions—to the biggest problems that we face as individuals, communities, and nations.

Key 42: The greatest crisis we face as individuals and as a society is a crisis of thinking. How can we make room for the new world that's emerging if we are clinging to the old world of the past?

Key 43: A growing body of scientific evidence is leading to an inescapable conclusion: Violent competition and war directly contradict our deepest instincts for cooperation and nurturing.

Key 44: The brutality of hate-based crimes is possible only in a society where the value of human life has been lost.

Key 45: The destruction of an individual through the abuse of drugs and alcohol is only possible when the personal sense of worth is lost.

Key 46: Rachel Carson reminds us that we only destroy what we don't value and we can't value what we don't know. A lasting solution to the issues that divide us and the growing levels of bullying, hate crimes, and wartime atrocities is to instill in the new generation, and embrace within ourselves, the need to respect and value all life.

RESOURCES

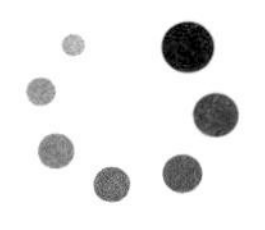

Heart Intelligence / Resilience

The Institute of HeartMath, www.HeartMath.org

"The Institute of HeartMath is an internationally recognized non-profit research and education organization dedicated to helping people reduce stress, self regulate emotions and build energy and resilience for happy, healthy lives. HeartMath tools, technology and training teach people to rely upon the intelligence of their hearts in concert with their minds at home, school, work and play."

Hate Crimes

National Organization for Victim Assistance (NOVA), www.trynova.org

Hate crimes create a complex set of circumstances and needs that vary from individual to individual. A number of U.S. states offer assistance to victims, as well as the training to professionals to understand how to address hate. This website is a portal to many of those organizations nationwide.

Recommended Reading

Charles Darwin. *On the Origin of Species by Means of Natural Selection* (Seattle, WA: Pacific Publishing Studio, 2010).

Doc Lew Childre, Howard Martin, and Donna Beech, *The HeartMath Solution: The Institute of HeartMath's Revolutionary Program for Engaging the Power of the Heart's Intelligence* (New York: HarperOne, 2000).

Francis Crick. *Life Itself: Its Origin and Nature* (New York: Touchstone, 1981).

Adrián Recinos. *Popol Vuh: The Sacred Book of the Ancient Quiché Maya,* Part I, Creation Myth, Chapters 1–3, Delia Goetz and Sylvanus G. Morley, editors (Norman, OK: University of Oklahoma Press, 1950). Available at: https://en.wikipedia.org/wiki/Popol_Vuh#Creation_myth.

ENDNOTES

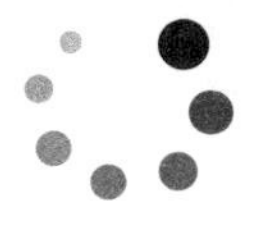

Epigraph. Carl Sagan. *Contact* (New York: Simon and Schuster, 1997), p. 430.

Introduction

1. There has been an explosion of new research exploring the power of human beliefs, the placebo effect, and the power of our expectations when it comes to the healing of the body. This particular example describes a randomized double-blind study that was done with a group experiencing Parkinson's disease. Joseph Mercola. "How the Power of Your Mind Can Influence Your Healing and Recovery," Mercola.com (March 5, 2015). Available at: http://articles.mercola.com/sites/articles/archive/2015/03/05/placebo-effect-healing-recovery.aspx.
2. Elizabeth Palermo, associate editor. "Niels Bohr: Biography & Atomic Theory," (May 14, 2013). Available at: http://www.livescience.com/32016-niels-bohr-atomic-theory.html.

Chapter 1 Breaking Darwin's Spell

Epigraph. Scott Turow. *Ordinary Heroes* (New York: Grand Central Publishing, 2011), p. 66.

1. Frank Newport. "In U.S., 42% Believe Creationist View of Human Origins," Gallup.com (June 2, 2014). Available at: http://www.gallup.com/poll/170822/believe-creationist-view-human-origins.aspx.
2. Francis Crick. *Life Itself: Its Origin and Nature* (New York: Touchstone, 1981), p. 88.
3. Adrián Recinos. *Popol Vuh: The Sacred Book of the Ancient Quiché Maya*, "Creation Myth," chapters 1–3, Delia Goetz and Sylvanus G. Morley, eds. (Norman, OK: University of Oklahoma Press, 1950), pp. 167–168. Available at: https://en.wikipedia.org/wiki/Popol_Vuh#Creation_myth. The P*opol Vuh*, as we know it today, is taken from the records of Francisco Ximénez, a Dominican priest, written at the turn of the 18th century. The manuscript sank into obscurity until it was "rediscovered" in 1941 by Adrián Recinos, who is generally given credit for publishing it in recent times. Recinos

explains: "The original manuscript is not divided into parts or chapters; the text runs without interruption from the beginning until the end. In this translation I have followed the Brasseur de Bourbourg division into four parts, and each part into chapters, because the arrangement seems logical and conforms to the meaning and subject matter of the work. Since the version of the French Abbe is the best known, this will facilitate the work of those readers who may wish to make a comparative study of the various translations of the Popol Vuh" (Goetz xiv; Recinos 11–12; Brasseur, xv).

4. *The Holy Bible: Authorized King James Version*, Genesis, chapter 1, verse 26 (Cleveland, OH: World Publishing Company, 1961), p. 9.

5. *The Torah: A Modern Commentary*, Bereshit, chapter 1, verse 26, W. Gunther Plaut, eds. (New York: Union of American Hebrew Congregations, 1981), p. 19.

6. "Ancient Egypt: The Mythology," EgyptianMyths.net. Available at: http://www.egyptianmyths.net/section-deities.htm.

7. These slogans (from Lucky Strike cigarettes, actor Edmund Lowe's endorsement of Lucky Strike, and Viceroy cigarettes) were popular in tobacco ads in the early to mid-20th century. See Hadgirl. "10 Evil Vintage Cigarette Ads Promising Better Health," Healthcare Administration Degree Programs blog. Available at: http://www.healthcare-administration-degree.net/10-evil-vintage-cigarette-ads-promising-better-health.

8. Ibid.

9. NBC TV news report (January 11, 1964) by correspondent Frank McGee. "Special Report: Smoking and Health." Available at: https://highered.nbclearn.com/portal/site/HigherEd/flatview?cuecard=68341.

10. Terry Pratchett. *A Hat Full of Sky* (New York: HarperCollins, 2004). Read excerpts from the book at: https://theillustratedpage.wordpress.com/2015/07/16/review-of-a-hat-full-of-sky-by-terry-pratchett.

11. Carl Sagan, "The Backbone of Night," *Cosmos* episode 7, November 9, 1980.

12. Albert Einstein, as cited by Steven Pollock, Oliver DeWolfe, and Steve Goldhaber, Physics Department, University of Colorado, Boulder. "Physics 3220: Quantum Mechanics" (Fall 2008). Available at: http://www.colorado.edu/physics/phys3220/phys3220_fa08/quotes.html.

13. Charles Darwin. *On the Origin of Species by Means of Natural Selection*. Available at: http://www.gutenberg.org/files/2009/2009-h/2009-h.htm.

14. For more information about Charles Darwin's voyage aboard the HMS *Beagle*, please see this website: https://www.aboutdarwin.com/voyage/voyage03.html.

15. Darwin, *Origin of Species*, pp. 126–7.

16. Ibid., p. 219.

17. Ibid., p. 155.

18. "*Evolution* Series Overview," PBS.org (2001). Available at: http://www.pbs.org/wgbh/evolution/about/overview.html.

19. Joshua Gilder. "PBS' 'Evolution' Series Is Propaganda, Not Science," WorldNetDaily.com (September 24, 2001). Available at: http://www.wnd.com/2001/09/11004.

20. Read the text of Oklahoma Senate Bill 1322, proposed by State Senator Josh Brecheen during the second session of the 55th Oklahoma State Legislature (2016) at: http://www.oklegislature.gov/BillInfo.aspx?Bill=sb1322&Session=1600.

21. "Definition of Intelligent Design," Discovery Institute, Center for Science and Culture website (accessed January 30, 2017). Available at: http://www.intelligentdesign.org/whatisid.php.

22. Decision filed December 20, 2005, in the Dover case by the United States District Court for the Middle District of Pennsylvania. "Tammy Kitzmiller, et al., v. Dover Area School District, et al.," National Center for Science Education website. Available at: https://ncse.com/files/pub/legal/kitzmiller/highlights/2005-12-20_Kitzmiller_decision.pdf.

23. Louis Agassiz. "Evolution and Permanence of Type," *Atlantic Monthly* (January 1874), p. 10. Available at: http://www.unz.org/Pub/AtlanticMonthly-1874jan-00092.

24. Ibid., p. 12, italics added.

25. Adam Sedgwick. *Spectator* (March 1860). Quoted in David L. Hull, *Darwin and His Critics: The Reception of Darwin's Theory of Evolution by the Scientific Community* (Cambridge, MA: Harvard University Press, 1973), pp. 155–170.

26. *Louis Agassiz: His Life and Correspondence*, Elizabeth C. Agassiz, eds. (Boston: Houghton Mifflin, 1893), p. 647. Available at: https://ia902606.us.archive.org/28/items/louisagassizhisl02agas/louisagassizhisl02agas.pdf.

27. Albert Fleischmann. "The Doctrine of Organic Evolution in the Light of Modern Research," *Journal of the Transactions of the Victoria Institute or Philosophical Society of Great Britain*, vol. 65 (London, U.K., 1933), pp. 194–195, 205–206, 208–9. Available at: https://biblicalstudies.org.uk/pdf/jtvi/1933_194.pdf.

28. H. S. Lipson. "A Physicist Looks at Evolution," *Physics Bulletin*, vol. 31, no. 4 (May 1980), p. 138.

29. Leonard Harrison Matthews. "Introduction," *The Origin of the Species* by Charles Darwin (London: J. M. Dent and Sons, 1971), pp. x–xi.

30. Fred Hoyle. "Hoyle on Evolution," *Nature*, vol. 294, no. 5837 (November 12, 1981), p. 105.

31. Michael Denton. *Evolution: A Theory in Crisis* (Chevy Chase, MD: Adler and Adler Books, 1986), p. 358.

32. Stephen Jay Gould. "Not Necessarily a Wing," *Natural History*, vol. 94, no. 14 (October 1985), pp. 12–13.

33. Wolfgang Smith. *Teilhardism and the New Religion: A Thorough Analysis of the Teachings of Pierre Teilhard de Chardin* (Charlotte, NC: TAN Books, 1988), p. 24.

34. "A Scientific Dissent from Darwin" is a website that contains the list launched by the Discovery Institute in 2001 of scientists throughout the world who have not accepted Darwin's theory of evolution as fact. Available at: http://www.dissentfromdarwin.org.

35. Charles Darwin to Asa Gray, 1860. Quoted in David Masci, "Darwin and His Theory of Evolution," Pew Research Center, Religion and Public Life (February 4, 2009). Available at: http://www.pewforum.org/2009/02/04/darwin-and-his-theory-of-evolution.

36. Henry Edward Manning. Quoted in Masci, "Darwin and His Theory of Evolution."

37. Thomas H. Morgan. *Evolution and Adaptation* (New York: Macmillan Company, 1903), p. 43.

38. Darwin (Pacific Publishing Studio, 2010), p. 151.

Chapter 2 Human by Design

Epigraph. Harold Urey, as cited by *Christian Science Monitor* (January 4, 1962), p. 4.

1. "This Day in History: February 28: Lead Story: Watson and Crick Discover Chemical Structure of DNA," History.com (accessed January 30, 2017). Available at: http://www.history.com/this-day-in-history/watson-and-crick-discover-chemical-structure-of-dna.

2. William Goodwin. "Rare Tests on Neanderthal Infant Sheds Light on Early Human Development," *Science News* (April 4, 2000). Available at: https://www.sciencedaily.com/releases/2000/03/000331091126.htm.

3. "What Does It Mean to Be Human? Neanderthal Mitochondrial DNA," Smithsonian Institution, National Museum of Natural History website (accessed January 30, 2017). Available at: http://humanorigins.si.edu/evidence/genetics/ancient-dna-and-neanderthals/neanderthal-mitochondrial-dna.

4. Igor V. Ovchinnikov, Anders Götherström, Galina P. Romanova, Vitaliy M. Kharitonov, Kerstin Lidén, and William Goodwin. "Molecular Analysis of Neanderthal DNA from the Northern Caucasus," *Nature,* vol. 404 (2000), pp. 490–493. Available at: http://cogweb.ucla.edu/Abstracts/Goodwin_00.html.

5. "What Does It Mean to Be Human? Homo Sapiens," Smithsonian Institution, National Museum of Natural History website (Accessed January 30, 2017). Available at: http://humanorigins.si.edu/evidence/human-fossils/species/homo-sapiens.

6. Lizzie Wade. "Oldest Human Genome Reveals When Our Ancestors Had Sex with Neandertals," *Science* website (October 22, 2014). Available at: http://www.sciencemag.org/news/2014/10/oldest-human-genome-reveals-when-our-ancestors-had-sex-neandertals.

7. Hillary Maywell. "Neandertals Not Our Ancestors, DNA Study Suggests," *National Geographic News* (May 14, 2003). Available at: http://news.nationalgeographic.com/news/2003/05/0514_030514_neandertalDNA.html.

8. Public Library of Science. "Europe's Ancestors: Cro-Magnon 28,000 Years Old Had DNA Like Modern Humans," *ScienceDaily* (July 16, 2008). Available at: www.sciencedaily.com/releases/2008/07/080715204741.htm.

9. Simon Tripp and Martin Grueber. "Economic Impact of the Human Genome Project," Battelle Memorial Institute report (May 2011). Available at: http://www.battelle.org/docs/default-document-library/economic_impact_of_the_human_genome_project.pdf.

10. For an easy-to-understand description of the DNA differences between humans and our nearest primate relatives, chimpanzees, visit "DNA: Comparing Humans and Chimps," American Museum of Natural History website (accessed January 30, 2017): http://www.amnh.org/exhibitions/permanent-exhibitions/human-origins-and-cultural-halls/anne-and-bernard-spitzer-hall-of-human-origins/understanding-our-past/dna-comparing-humans-and-chimps.

11. The term *7q31* is shorthand notation for the way scientists describe the location of a gene within a chromosome. The code is simple and made of three parts. Part 1: The first number tells us the big picture of which chromosome the gene is located within. Part 2: The letter tells us which of the two arms that make up a chromosome the gene is on: the short (or p) arm, or the long (or q) arm. Part 3: The latter number tells us the gene's actual position on the chromosome, as determined by the number of dark and light bands that are visible using a microscope on specially stained samples. In this case the gene is on chromosome 7, on the long q arm, at position 31, when counting from the center point (centromere) of the chromosome.

12. "Study Links Evolution of Single Gene to Human Capacity for Language," Emory University, Yerkes National Primate Research Center press release (November 11, 2009). Available at: http://www.yerkes.emory.edu/about/news/neuropharmacology_neurologic_diseases/gene_language_capacity.html.

13. Ibid.

14. Wolfgang Enard, interviewed by Helen Briggs. "First Language Gene Discovered," *BBC News* World Edition (August 14, 2002). Available at: http://news.bbc.co.uk/2/hi/science/nature/2192969.stm.

15. Ibid.

16. Michael Purdy. "Human Chromosomes 2, 4 Include Gene Deserts, Signs of Chimp Chromosome Merger," *Washington University in St. Louis Source* (April 6, 2005). Available at: https://source.wustl.edu/2005/04/human-chromosomes-2-4-include-gene-deserts-signs-of-chimp-chromosome-merger. See also J. W. Ijdo, A. Baldini, D. C. Ward, S. T. Reeders, and R. A. Wells. "Origin of Human Chromosome 2: An Ancestral Telomere-Telomere Fusion," *Proceedings of the National Academy of Sciences USA*, vol. 88, no. 20 (October 15, 1991), pp. 9051–5. Available at: https://www.ncbi.nlm.nih.gov/pmc/articles/PMC52649.

17. J. W. Ijdo et al. Italics added. While some scientists continue to object to the conclusion that human chromosome 2 is the result of an ancient gene

fusion, the evidence clearly points to such a fusion. To summarize, the evidence states that: (1) the fact that the DNA sequences of the separate chimp genes are nearly identical to those found combined in human chromosome 2; (2) the presence of a second unused "vestigial" centromere (the point that separates the long and short arms of the gene), which would be expected if two genes, each with one centromere, had merged into a single unit; and (3) the presence of vestigial telomeres—the protective sequence of DNA normally found at the ends of chromosomes— found midgene in the q13 band, rather than at the end of the chromosome.

18. For a detailed description of functions associated with human chromosome 2, visit "Chromosome 2 (Human)," Wikipedia (accessed January 30, 2017). Available at: https://en.wikipedia.org/wiki/Chromosome_2_(human).
19. Ibid.
20. J. W. Ijdo, et al.
21. *The Expanded Quotable Einstein*, Alice Calaprice, ed. (Princeton, NJ: Princeton University Press, 2000), p. 204.
22. Alfred Russel Wallace. *Contributions to the Theory of Natural Selection* (New York: Macmillan, 1870), p. 356. Available at: https://ia601406.us.archive.org/32/items/contributionstot00wall/contributionstot00wall.pdf.

Chapter 3 The Brain in the Heart

Epigraph. Gary E. R. Schwartz and Linda G. S. Russek. Foreword to Paul P. Pearsall, *The Heart's Code: Tapping the Wisdom and Power of Our Heart Energy* (New York: Broadway Books, 1998), p. xiii.

1. "Cro-Magnon," Wikipedia (accessed January 30, 2017). Available at: https://en.wikipedia.org/wiki/Cro-Magnon.
2. Ibid.
3. "Neanderthal Anatomy," Wikipedia (accessed January 30, 2017). Available at: https://en.wikipedia.org/wiki/Neanderthal_anatomy.
4. Joshua Batson. "Watch 80,000 Neurons Fire in the Brain of a Fish," *Wired* (July 28, 2014). Available at: https://www.wired.com/2014/07/neuron-zebrafish-movie.
5. "Anatomy of the Brain," Mayfield Clinic, Brain and Spine Institute (accessed January 30, 2017). Available at: http://www.mayfieldclinic.com/PE-AnatBrain.htm#.VYTaBFVViko.
6. "Amazing Heart Facts," Arkansas Heart Hospital (accessed January 30, 2017). Available at: http://www.arheart.com/cardiovascular-health/amazing-heart-facts.
7. The languages of Hebrew, Aramaic, and ancient Greek have contributed to the Bible we know today. When its passages are translated into English, the exact number of times a particular word occurs in the Bible varies depending upon the translation (for example, the Authorized King James Version or the New American Standard). To learn the number of times the word *heart* occurs in different versions of the Bible, see "Word Counts: How Many Times

Does a Word Appear in the Bible?" Christian Bible Reference Site: http://www.christianbiblereference.org/faq_WordCount.htm.

8. *The Holy Bible, Authorized King James Version*, Proverbs, chapter 20, verse 5 (Cleveland, OH: World Publishing Company, 1961), p. 534.
9. Rodney Ohebsion. "Native American Proverbs, Quotes and Chants," RodneyOhebsion.com (accessed January 30, 2017). Available at: http://www.rodneyohebsion.com/native-american-proverbs-quotes.htm.
10. Daisaku Ikeda. "The Wisdom of the Lotus Sutra," Soka Gakki International (accessed January 30, 2017). Available at: http://www.sgi.org/about-us/president-ikedas-writings /the-wisdom-of-the-lotus-sutra.html.
11. Ibid.
12. See Ralph Marinelli, Branko Fuerst, Hoyte van der Zee, Andrew McGinn, and William Marinelli. "The Heart Is Not a Pump," *Frontier Perspectives* (Fall/Winter 1995). Available at: http://www.rsarchive.org/RelArtic/Marinelli.
13. J. Andrew Armour. *Neurocardiology: Anatomical and Functional Principles*, HeartMath Research Center, Institute of HeartMath, eBook (2003).
14. Ibid.
15. Ibid.
16. Ibid.
17. The Quick Coherence® Technique for Adults. Available at: https://www.heartmath.org/resources/heartmath-tools/quick-coherence-technique-for-adults.
18. Armour. *Neurocardiology.*
19. "Fifty Spiritual Homilies of Saint Macarius the Egyptian: Homily 43:7," e-Catholic 2000 (accessed March 22, 2017). Available at: http://www.ecatholic2000.com/macarius/untitled-46.shtml#_Toc385610658.
20. Tony Long. "Dec. 3, 1967: Patient Dies, but First Heart Transplant a Success," *Wired* (December 3, 2007). Available at: https://www.wired.com/2007/12/dayintech-1203.
21. "Artificial Hearts May Help Patients Survive until Transplant," American College of Cardiology press release (March 27, 2014). Available at: http://www.acc.org/about-acc/press-releases/2014/03/27/12/53/gurudevan-artificial-heart-pr.
22. Ibid.
23. Claire Sylvia. *A Change of Heart: A Memoir* (New York: Warner Books, 1997).
24. Ibid., p. 226.
25. Paul Pearsall. *The Heart's Code* (New York: Broadway Books, 1999), Introduction.
26. Charles E. Gross. "Leonardo da Vinci on the Brain and the Eye," *Neuroscientist*, vol. 3, no. 5 (September 1, 1997), pp. 347–54. Available at: http://journals.sagepub.com/doi/pdf/10.1177/107385849700300516.

27. Clare Boothe Brokaw (Clare Boothe Luce). *Stuffed Shirts* (New York, Horace Liveright, 1931), p. 239.

28. Chad Boutin. "Snap judgments decide a face's character, psychologist finds," Princeton University (August 22, 2006). Available at: https://www.princeton.edu/main/news/archive/S15/62/69K40/index.xml?section=topstories.

29. My affiliation with the Institute of HeartMath dates to 1995. During that time I've shared keynote presentations and weekend seminars with Howard Martin, executive vice president, and Debbie Rozman, Ph.D., president and co-CEO; and I have served on the steering committee for the Global Coherence Initiative Project since its inception in 2008. For a list of staff and advisors, go to: https://www.heartmath.com/heartmath-team.

30. Rollin McCraty, Mike Atkinson, and Raymond Trevor Bradley. "Electrophysiological Evidence of Intuition: Part 1. The Surprising Role of the Heart," *Journal of Alternative and Complementary Medicine*, vol. 10, no. 1 (June 2004), pp. 133–143.

Chapter 4 The New Human Story

Epigraph. Brené Brown. *Own Our History. Change the Story* (June 18, 2015). Available at: http://brenebrown.com/2015/06/18/own-our-history-change-the-story.

1. Kristen Philipkoski. "Researchers Cut Gene Estimate," *Wired* (February 12, 2001). Available at: http://archive.wired.com/science/discoveries/news/2001/02/41749.

2. "The Human Genome Is More and Less Than We Expected to Find," The Tech Museum of Innovation (2013). Available at: http://genetics.thetech.org/original_news/news14.

3. Guilherme Neves, Jacob Zucker, Mark Daly, and Andrew Chess. "Stochastic Yet Biased Expression of Multiple *Dscam* Splice Variants by Individual Cells," *Nature Genetics*, vol. 36, no. 3 (February 1, 2004), pp. 240–6.

4. Victor A. McKusick. As quoted in "2001: Publication of the Human Genome Sequence," *Genome News Network*. Available at: http://www.genomenewsnetwork.org/resources/timeline/2001_human_pub.php.

5. Craig Venter. As quoted by Tom Abate. "Genome Discovery Shocks Scientists," *San Francisco Chronicle* (February 11, 2001). Available at: http://www.sfgate.com/news/article/Genome-Discovery-Shocks-Scientists-Genetic-2953173.php.

6. Albert A. Michelson and Edward W. Morley. "On the Relative Motion of the Earth and the Luminiferous Ether," *American Journal of Science*, vol. 34, no. 203 (November 1887), pp. 333–45.

7. E. W. Silvertooth. "Special Relativity." *Nature*, vol. 322, no. 6080 (August 1986), p. 590.

8. Ilya Prigogine, Gregoire Nicolis, and Agnes Babloyantz. "Thermodynamics of Evolution," *Physics Today*, vol. 25, no. 11 (November 1972), pp. 23–8.

9. Marcel Golay and Frank Salisbury. As quoted by Henry M. Morris. "Probability and Order versus Evolution," *Acts and Facts*, vol. 8, no. 7 (1979). Available at: http://www.icr.org/article/probability-order-versus-evolution.

10. Fred Hoyle and N. Chandra Wickramasinghe. *Evolution from Space* (London: J. M. Dent & Sons, 1981).

11. Fred Hoyle. "Hoyle on Evolution," *Nature*, vol. 294, no. 5837 (November 12, 1981), p. 105.

12. John Black. "The Origins of Human Beings according to Ancient Sumerian Texts," *Ancient Origins* (January 30, 2013). Available at: http://www.ancient-origins.net/human-origins-folklore /origins-human-beings-according-ancient-sumerian-texts-0065.

13. Louis Ginzberg. *The Legends of the Jews*, vol. 1, *From Creation to Jaco* (1938), p. 54. Available at: http://www.gutenberg.org/ebooks/1493.

14. *The Holy Qur'an, with English Translation and Commentary*, Pilgrimage, chapter 22, verse 5. Maulana Muhammad Ali, ed. (Columbus, OH: Ahmadiyah Anjuman Isha'at Islam, 1917), p. 648.

15. Ibid., chapter 25, verse 54, p. 705.

16. Ibid., p. 648.

17. *The Holy Bible, Authorized King James Version*, Genesis, chapter 2, verse 7 (Cleveland, OH: World Publishing Company, 1961), p. 10.

18. Charles C. Mann. *1491: New Revelations of the Americas before Columbus* (New York: Alfred A. Knopf, 2005), pp. 199–212.

19. *Popol Vuh*, Norine Polio, ed., Yale-New Haven Teachers Institute. Available at: http://teachersinstitute.yale.edu/curriculum/units/1999/2/99.02.09.x.html.

20. Duane Elgin. "Why We Need to Believe in a Living Universe," *Huffington Post* blog (May 15, 2011). Available at: http://www.huffingtonpost.com/duane elgin/living-universe_b_862220.html.

21. Ibid.

22. Ibid.

23. Ibid.

24. Ray Bradbury. "G. B. S. Mark V," in *I Sing the Body Electric! And Other Stories* (New York: HarperPerennial, 2001), p. 275.

25. Albert Einstein. Letter to Robert S. Marcus, political director of the World Jewish Congress, on the occasion of his son passing away from polio (February 12, 1950), emphasis added.

26. Karl Jaspers. *The Idea of the University* (London: Peter Owen, 1965), p. 30. As cited by James Cowan. "Climate Change: A Humanist Response," epigraph (June 2015). Available at: http://www.academia.edu/12372530 /Climate_Change_a_humanist_response.

Chapter 5 We're "Wired" for Connection

Epigraph. Mitch Albom. *The Five People You Meet in Heaven* (New York: Hachette, 2003), p. 50.

1. Dean Koontz, as cited on Goodreads. Available at: http://www.goodreads.com/quotes/95562-intuition-is-seeing-with-the-soul.
2. "Mother-Baby Study Supports Heart-Brain Interactions," HeartMath Institute (April 20, 2008). Available at: https://www.heartmath.org/articles-of-the-heart/science-of-the-heart/mother-baby-study-supports-heart-brain-interactions.
3. Ibid.
4. Ibid.
5. Ibid.
6. "Captured Pilot's Mother Felt Something Was Wrong," CNN.com (March 24, 2003). Available at: http://www.cnn.com/2003/US/South/03/24/sprj.irq.pilot.family.
7. Ibid.
8. Ibid.
9. Alan Cowell and Douglas Jehl. "Luxor Survivors Say Killers Fired Methodically," *New York Times* (November 24, 1997). Website: http://www.nytimes.com/1997/11/24/world/luxor-survivors-say-killers-fired-methodically.html.
10. Albert Einstein. Letter to Robert S. Marcus (February 12, 1950).
11. The Dalai Lama. *The Art of Happiness: A Handbook for Living*, 10th anniversary edition (New York: Riverhead Books, 2009), p. 119.
12. Joanna Macy, "The Bodhisattva," excerpted from a talk at Barre Center for Buddhist Studies, "The Wings of the Bodhisattva," *Insight Magazine* (Spring/Summer 2001). Available at: http://www.joannamacy.net/the-bodhisattva.html.

Chapter 6 We're "Wired" for Healing and Long Life

Epigraph. Neel Burton. Available at: http://www.goodreads.com/quotes/7280473-many-things-can-prolong-your-life-but-only-wisdom-can.

1. *The Holy Bible, Authorized King James Version*, Genesis, chapter 6, verse 10 (Cleveland, OH: World Publishing Company, 1961), p. 13.
2. Ibid., Genesis, chapter 5, verse 24, p. 12.
3. Ibid., Genesis, chapter 6, verse 3, p. 13.
4. "The Nobel Prize in Physiology or Medicine 2009," Nobelprize.org press release (October 5, 2009). Available at: https://www.nobelprize.org/nobel_prizes/medicine/laureates/2009/press.html.

5. Ewen Callaway. "Telomerase Reverses Aging Process," *Nature News* (November 28, 2010). Available at: http://www.nature.com /news/2010/101128/full/news.2010.635.html

6. Ibid.

7. Kristin Kirkpatrick. "Should I Stop Eating Eggs to Control Cholesterol? (Diet Myth 4)," ClevelandClinic.org (August 16, 2012). Available at: https://health. clevelandclinic.org/2012/08 /should-i-stop-eating-eggs-to-control-cholesterol-diet-myth-4.

8. John Phillip. "Targeted Nutrients Naturally Extend Telomere Length and Provide Anti-aging Effect," *Natural News* (December 29, 2011). Available at: http://www.naturalnews.com/034513_telomeres_longevity_nutrition.html. Original study available at http://jn.nutrition.org/content/139/7/1273.full .pdf.

9. Elissa S. Epel, Elizabeth H. Blackburn, Jue Lin, Firdaus S. Dhabhar, Nancy E. Adler, Jason D. Morrow, and Richard M. Cawthon. "Accelerated Telomere Shortening in Response to Life Stress," *Proceedings of the National Academy of Sciences of the United States of America*, vol. 101, no. 49 (September 28, 2004), pp. 17312–5. Available at: http://www.pnas.org/content/101/49/17312.long.

10. "Essenes," Wikipedia (accessed January 30, 2017). Available at: https:// en.wikipedia.org/wiki/Essenes.

11. *The Essene Gospel of Peace*, Edmond Bordeaux Szekely, ed. and trans. (Matsqui, BC: International Biogenic Society, 1937), p. 39.

12. There are legal and cultural definitions for precisely what *food* is. I'm using a Google definition that addresses the common and practical aspects of food as it's understood in our society. See https://www.google.com /webhp?sourceid=chrome-instant&ion=1&espv=2&ie=UTF -8#q=definition+of+food.

13. "Li Ching-Yuen," Wikipedia (accessed January 30, 2017). Available at: https://en.wikipedia.org/wiki/Li_Ching-Yuen.

14. "Li Ching-Yun Dead; Gave His Age as 197," *New York Times* (May 6, 1933); available at: http://query.nytimes.com/gst/abstract.html?res=9503E4DF153 8E333A25755C0A9639C946294D6CF; and "China: Tortoise-Pigeon-Dog," *Time* (May 15, 1933); available at: http://content.time.com/time/magazine /article/0,9171,745510,00.html.

15. "Tortoise-Pigeon-Dog," *Time*.

16. Martin Patience, "World's 'Oldest' Person in Israel," *BBC News* (February 15, 2008). Available at: http://news.bbc.co.uk/2/hi/middle_east/7247679.stm.

17. "Life Expectancy for Social Security," Social Security Administration (accessed January 30, 2017). Available at: https://www.ssa.gov/history /lifeexpect.html.

18. Romeo Vitelli. "When a Parent Loses a Child," *Psychology Today* (February 4, 2013). Available at: https://www.psychologytoday.com/blog /media-spotlight/201302/when-parent-loses-child.

19. American Psychological Association. "What Is Resilience?" Psych Central (accessed March 20, 2017). Available at: http://psychcentral.com/lib/2007/what-is-resilience

20. "What Is Resilience?" Stockholm Resilience Centre (July 4, 2008). Available at: http://www.stockholmresilience.org/research/research-videos/2011-12-01-what-is-resilience.html.

21. "Heart Rate Variability," HeartMath Institute (October 27, 2014). Available at: https://www.heartmath.org/articles-of-the-heart/the-math-of-heartmath/heart-rate-variability.

22. Rollin McCraty, Raymond Trevor Bradley, and Dana Tomasino. "The Resonant Heart," *Shift* (December 2004–February 2005), pp. 15–19. Available at: https://www.heartmath.org/research/research-library/relevant-publications/the-resonant-heart.

23. Doc Childre and Deborah Rozman. *Transforming Stress: The HeartMath Solution for Transforming Worry, Fatigue, and Tension* (Oakland, CA: New Harbinger Publications, 2005), p. 99.

Chapter 7 We're "Wired" for Destiny

Epigraph. William Jennings Bryan, from "America's Mission," a speech he delivered at a banquet given by the Virginia Democratic Association in Washington, DC, on February 22, 1899. Available at: https://archive.org/stream/speechesofwillia02bryauoft/speechesofwillia02bryauoft_djvu.txt.

1. *Forrest Gump* (1994), directed by Robert Zemeckis. Written by Eric Roth, based on the novel *Forrest Gump* by Winston Groom (New York: Vintage Books, 1986).

2. Aldous Huxley. *Brave New World* (London, U.K.: Chatto and Windus, 1931).

3. H. G. Wells. *Men Like Gods* (London, U.K.: Cassell & Company, 1921).

4. Carnegie Endowment for International Peace Records, 1910–1954, Carnegie Collections Rare Book and Manuscript Library, Columbia University. Available at: http://www.columbia.edu/cu/lweb/eresources/archives/rbml/CEIP/index.html?ceipFBio.html&1.

5. "11 Myths about Global Hunger," World Food Programme (October 21, 2011). Available at: https://www.wfp.org/stories/11-myths-about-global-hunger.

6. "By the Numbers: Hunger in the World," UFCW Canada (United Food and Commercial Workers Union 2017), Available at: "http://www.ufcw.ca/index.php?option=com_content&view=article&id=3061:by-the-numbers-hunger-in-the-world&catid=271&Itemid=6&lang=en.

7. Richard Martin. "Meltdown-Proof Reactors Get a Safety Check in Europe," *MIT Technology Review* (September 4, 2015). Available at: https://www.technologyreview.com/s/540991/meltdown-proof-nuclear-reactors-get-a-safety-check-in-europe.

8. Ibid.

9. "Indian Point Energy Center," Wikipedia (accessed January 30, 2017). Available at: https://en.wikipedia.org/wiki/Indian_Point_Energy_Center.
10. Doug Stephens. "Shared Interests: The Rise of Collaborative Consumption," Retail Prophet (November 26, 2013). Available at: http://www.retailprophet.com/blog/shared-interests-the-rise-of-collaborative-consumption.
11. Stephen Hawking, from an interview published in a German magazine (translator unknown). Von Klaus Franke and Henry Glass. "Wir alle wollen wissen, woher wir kommen," *Der Spiegel,* vol. 42 (October 17, 1988). Available at: http://www.spiegel.de/spiegel/print/d-13542088.html.
12. Richard Dawkins. "Review of *Blueprints: Solving the Mystery of Evolution,*" *New York Times* (April 9, 1989), p. 34.
13. Neil Munro. "Poll: Race Relations Have Plummeted Since Obama Took Office," *Daily Caller* (July 25, 2013). Available at: http://dailycaller.com/2013/07/25/race-relations-have-plummeted-since-obama-took-office-according-to-poll.
14. Eric Hobsbawm, "War and Peace in the 20th Century," *London Review of Books,* vol. 24, no. 4 (February 21, 2002). Hobsbawm's statistics show that by the end of the 20th century, over 187 million people had lost their lives to war. Available at: https://www.lrb.co.uk/v24/n04/eric-hobsbawm/war-and-peace-in-the-20th-century.
15. Matthew White. "Worldwide Statistics of Casualties, Massacres, Disasters and Atrocities," *The Historical Atlas of the Twentieth Century*. Available at: http://necrometrics.com/index.htm.
16. Jonathan Steele. "The Century That Murdered Peace," *Guardian* (December 11, 1999). Available at: https://www.theguardian.com/world/1999/dec/12/theobserver4.
17. "Convention on the Prevention and Punishment of the Crime of Genocide," UN General Assembly resolution (December 9, 1948). Available at: http://www.ohchr.org/EN/ProfessionalInterest/Pages/CrimeOfGenocide.aspx.
18. Richard Weikart. *From Darwin to Hitler: Evolutionary Ethics, Eugenics and Racism in Germany* (New York: Macmillan, 2006).
19. See Stéphane Courtois, Nicolas Werth, Jean-Louis Panné, Andrzej Paczkowski, Karel Bartošek, and Jean-Louis Margolin. *The Black Book of Communism.* Jonathan Murphy and Mark Kramer, trans. (Cambridge, MA: Harvard University Press, 1999), p. 491.
20. See Adolf Hitler. "Nation and Race," *Mein Kampf,* vol. 1: *A Reckoning* (1925). Available at: http://www.hitler.org/writings/Mein_Kampf/mkv1ch11.html.
21. "Past Genocides and Mass Atrocities." United to End Genocide. Available at: http://endgenocide.org/learn/past-genocides.
22. Charles Darwin, *On the Origin of Species by Means of Natural Selection* (Seattle: Pacific Publishing Studio, 2010), p. 133.
23. Hitler, *Mein Kampf.*
24. Charles Darwin. *The Descent of Man* (Amherst, NY: Prometheus Books, 1998), p. 110.

25. Peter Kropotkin. *Mutual Aid: A Factor of Evolution* (1902) (Boston: Porter Sargent, 1976), p. 14.

26. John M. Swomley. "Violence: Competition or Cooperation," *Christian Ethics Today*, vol. 26 (February 2000), p. 20. Available at: http://pastarticles.christianethicstoday.com/cetart/index.cfm?fuseaction=Articles.main&ArtID=300.

27. Ibid.

28. Ibid. Quoted in Ronald Logan. "Opening Address of the Symposium on the Humanistic Aspects of Regional Development," *Prout Journal*, vol. 6, no. 3 (September 1993).

29. Alfie Kohn. Quoted in Ronald Logan, "Opening Address."

30. Carl Sandburg. *The People, Yes* (1936) (New York: Mariner Books, 1990), p. 43.

31. Simone Robers, Anlan Zhang, Rachel E. Morgan, and Lauren Musu-Gillette. *Indicators of School Crime and Safety: 2014*, a report by the National Center for Education Statistics, Institute of Education Sciences (July 2015). Available at: https://nces.ed.gov/pubs2015/2015072.pdf.

32. "Suicide of Jadin Bell," Wikipedia (accessed January 30, 2017). Available at: https://en.wikipedia.org/wiki/Suicide_of_Jadin_Bell.

33. Ibid.

34. "Cyberbullying and Social Media," Megan Meier Foundation (accessed March 21, 2017). Available at: http://www.meganmeierfoundation.org/cyberbullying-social-media.html; Joe Vallese. "'Audry and Daisy' Exposes the Trauma of Teenage Sexual Assault and Slut Shaming," Vice (September 23, 2016). Available at: https://www.vice.com/en_us/article/audrie-and-daisy-netflix-documentary-social-media-sexual-assault.

35. Haeyoun Park and Iaryna Mykhyalyshyn. "L.G.B.T. People Are More Likely to Be Targets of Hate Crimes Than Any Other Minority Group," *New York Times* (June 16, 2016). Available at: https://www.nytimes.com/interactive/2016/06/16/us/hate-crimes-against-lgbt.html.

36. "Matthew Shepard," Wikipedia (accessed January 30, 2017). Available at: https://en.wikipedia.org/wiki/Matthew_Shepard.

37. In addition to offering a factual account of the murder of James Byrd, Jr., the Wikipedia entry "James Byrd, Jr." (accessed on January 30, 2017) describes the federal legislation that resulted from his death and the death of Matthew Shepard, the Hate Crimes Prevention Act. Available at: https://en.wikipedia.org/wiki/Murder_of_James_Byrd_Jr.

38. A transcript of testimony presented to the British House of Commons by Deputy Speaker Natascha Engel. "DAESH: Genocide of Minorities," *House of Commons Hansard*, vol. 608 (April 20, 2016). Available at: https://hansard.parliament.uk/commons/2016-04-20/debates/16042036000001/DaeshGenocideOfMinorities.

39. Ibid.

40. "FBI Releases 2014 Hate Crime Statistics," FBI National Press Office, Washington, DC (November 16, 2015). Available at: https://www.fbi.gov/news/pressrel/press-releases/fbi-releases-2014-hate-crime-statistics.

41. Ibid.

42. Tara Lawley-Bergey. "'My Heart Died': A Sister Writes about Losing Her Brother to a Drug Overdose," NBC10 (February 8, 2016). Available at: http://www.nbcphiladelphia.com/news/local/My-Heart-Died-A-Sister-Writes-About-Losing-Her-Brother-to-a-Drug-Overdose-367969281.html.

43. Ibid.

44. Ibid.

45. Rachel Carson was a marine biologist and conservationist whose 1962 book, *Silent Spring* (New York: Houghton Mifflin), originally published as a series of articles in *The New Yorker*, catapulted the environmental movement into mainstream awareness and eventually led to a ban on pesticides such as DDT.

46. "Rose Schneiderman," Wikipedia (accessed January 30, 2017). Available at: https://en.wikipedia.org/wiki/Rose_Schneiderman.

47. "Domestic Violence Statistics," Hope Rising (accessed January 20, 2017). Available at: http://hoperisingtx.org/about/domestic-violence-statistics.

48. Jim Yardley. "Report on Deadly Factory Collapse in Bangladesh Finds Widespread Blame," *New York Times* (May 22, 2013). Available at: http://www.nytimes.com/2013/05/23/world/asia/report-on-bangladesh-building-collapse-finds-widespread-blame.html.

49. Albert Schweitzer. *Reverence for Life*, Reginald H. Fuller, trans. (New York: Harper and Row, 1969).

50. Ibid.

51. Desmond Tutu. "Made for Goodness," *Huffington Post* (March 13, 2012). Available at: http://www.huffingtonpost.com/desmond-tutu/made-for-goodness_b_1199864.html.

Chapter 8 Where Do We Go from Here?

Epigraph. Henry Miller. *Big Sur and the Oranges of Hieronymus Bosch* (New York: New Directions, 1957), p. 25.

1. The precise language of this statement, though commonly attributed to Chief Seattle, has recently come into question. While the words may vary, the essence of what is attributed to him is consistent with his thinking, as evidenced in the 1854 speech he is most widely known for. Speech with commentary by Walt Crowley available at: http://www.historylink.org/File/1427.

2. Kahlil Gibran. *The Prophet* (New York: Alfred A. Knopf, 1963), p. 28.

INDEX

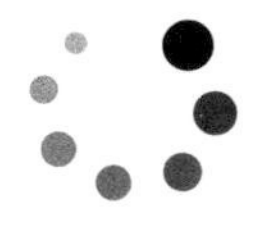

NOTE: Page references in *italics* refer to figures.

C

F

G

H

I

J

K

L

O

P

Q

R

S

T

U

V

W

Y

Z

ACKNOWLEDGMENTS

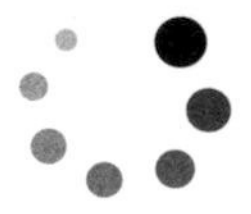

I remember the moment that I made the decision to write *The Science of Self-Empowerment.* I was returning home following three days of presentations at a conference in London. As I made my way past the television monitors along the airport concourse I noticed a common theme that wove each separate broadcast into the shared story that was washing across the airwaves that evening.

From the tragedies of domestic violence in the United States and the growing trend of cyberbullying among young people to the epidemic of illegal drug use throughout America and the unspeakable atrocities occurring in war-torn Syria and Iraq, the theme at the core of each news broadcast was the same: It was a human story based in the lack of value for human life. It was clear to me that any solution to ease such tragedy and suffering must address that core—*the fundamental way that we think of ourselves, and one another.* In that moment this book was conceived. I wanted to create a concise and accessible source of new discoveries that give us the reasons to change the way we think of ourselves. But a book is only an idea until it is given form.

If it takes a village to raise a child, it takes a community of like-minded individuals with diverse skills, spread across many time zones, to birth a book into being. This section is my opportunity to express my gratitude and appreciation to the family that has supported my commitment to share our new human story—the copy editors, proof editors, page layout designers, graphic designers, marketing representatives, publicists, and event producers who have worked behind the scenes to make this book possible. To each and every one of the most dedicated Hay House family I could ever imagine working with, I'm especially grateful to:

Louise Hay, Reid Tracy, and Margarete Nielsen—Thank you for your trust in me, your vision of how we as authors can contribute to our communities, and your dedication to the truly extraordinary way of doing business that has become the hallmark of Hay House's success.

Patty Gift—I'm so deeply grateful to you for believing in me from the beginning, for your ever-present support, your trust, and especially your friendship. *The Science of Self-Empowerment* marks my ninth book with Hay House and my 13-year anniversary as a Hay House author. I'm excited to see where the next 13 years lead us!

Anne Barthel—I'm grateful beyond words for your guidance, support, and friendship. The advice that you've shared with me extends far beyond the official title of executive editor that you hold, and is appreciated more that I could ever convey in words.

Richelle Fredson—You are a joy to work with and your publicity instincts are always spot-on. Thank you for your dedication to helping me reach as many people as possible with our empowering message and for making it so much fun to do so.

Christy Salinas and Tricia Breidenthal—You, and your extraordinarily talented staff, have been so patient with me, so open to my ideas, and so right on when it comes to the most beautiful book covers I could ever imagine that "thank you" seems inadequate to express the depth of my gratitude to you.

Kathryn Wells—Our awesome web project manager extraordinaire. I feel so very fortunate to have you and your team supporting me. My deepest gratitude to you for the most beautiful website, and the most inspiring newsletters I've ever had!

Mollie Langer—The best events producer I could ask for! Thank you for your dedication and professionalism, for honoring our audiences with the most beautiful live events on the planet, for the care that goes into everything you do, and especially for your friendship.

Rocky George—You're the perfect audio engineer with the ear for just the right sound always. I wish I could take you to each recording I do throughout the world!

Diane Ray and the entire Hay House Radio team—Thank you for making radio so fun and easy. My heartfelt gratitude to you for your dedication to excellence and for making me sound so good with each web cast, and every interview and radio show.

Melissa Brinkerhoff and all of the always-smiling, hard-working people who create the perfectly stocked book tables at our I Can Do It! and Celebrate Your Life events—you're all absolutely the very best! I couldn't ask for a more dedicated team, or a nicer group of people, to support my work. Your excitement and professionalism are unsurpassed, and I'm proud to be a part of all the good things that the Hay House family brings to our world.

Ned Leavitt—Thank you once again for your wisdom and the human touch that you bring to every project we share. I'm so deeply grateful to your guidance as my agent in the many and varied forms that it now involves, and I'm especially grateful for your trust in me, and our friendship.

Stephanie Gunning—My first-line editorial guru, sounding board, and now my friend of over 17 years. I'm deeply grateful for your wisdom, objectivity, and dedication to help me share the complexities of science and the truths of life in a joyous and meaningful way.

I am proud to be part of the virtual team, and the family, that has grown around the support of my work over the years, including my dearest Lauri Willmot, my favorite (and only) Executive Office Manager and the voice of Gregg Braden and Wisdom Traditions since 1996. I admire you tremendously, respect you deeply, and appreciate the countless ways that you're there for me always, every day and at all hours, your constant love and support, and especially your friendship.

Rita Curtis—I deeply appreciate your vision, your clarity as my business manager, and your skills that get us from here to there each month. I appreciate your trust, your openness to new ideas, and especially your friendship.

To my mother, the beautiful Sylvia Lee Braden—You fought for my life when I was in your womb, and now I have the honor of advocating for your health and dignity as your life is changing faster than either of us could have imagined. To my brother, Eric, my deepest gratitude for your unfailing love, and for believing in me even when you don't understand me. Though our family is small, together we have found that our extended family of love is greater than we've ever imagined.

Martha—My beautiful wife and best friend. Thank you beyond words for your acceptance and support, your unwavering friendship, your exquisite and gentle wisdom and all-embracing love that is with me each day of my life. Along with Woody, Nemo, and Mr. Merlin, the creatures that we share our lives with, you are the family that makes each journey worth coming home from. Thank you for all that you give, all that you share, and all of the joy that you bring to my life.

A very special "thank you" to everyone who has supported my work, books, recordings, and live presentations over the years. I am honored by your trust, in awe of your vision for a better world, and deeply appreciative of your passion to bring that world into existence. Through you, I have learned to become a better listener, and heard the words that allow me to share our empowering message of hope and possibility. To all, I remain grateful in all ways, always.

ABOUT THE AUTHOR

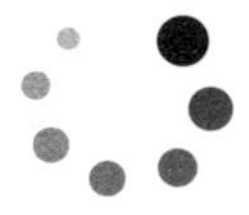

Gregg Braden is a five-time *New York Times* best-selling author and is internationally renowned as a pioneer in bridging science, spirituality, and the real world. From 1979 to 1990 Gregg worked for Fortune 500 companies such as Cisco Systems, Philips Petroleum, and Martin Marietta Defense systems as a problem solver during times of crisis. He continues problem-solving today as he weaves modern science and the wisdom preserved in remote monasteries and forgotten texts into real world solutions. His discoveries have led to 11 award-winning books now published in over 40 languages. The United Kingdom's *Watkins Journal* lists Gregg among the top 100 of "the world's most spiritually influential living people" for the fifth consecutive year, and in 2017 he received a nomination for the prestigious Templeton Award. He's shared his presentations and trainings with the United Nations, Fortune 500 companies, and the U.S. military and is now featured on major networks throughout North and South America, Mexico, and Europe.

Hay House Titles of Related Interest

YOU CAN HEAL YOUR LIFE, the movie, starring Louise Hay & Friends
(available as a 1-DVD program, an expanded 2-DVD set,
and an online streaming video)
Learn more at www.hayhouse.com/louise-movie

THE SHIFT, the movie,
starring Dr. Wayne W. Dyer
(available as a 1-DVD program, an expanded 2-DVD set,
and an online streaming video)
Learn more at www.hayhouse.com/the-shift-movie

BECOMING SUPERNATURAL: How Common People Are Doing the Uncommon, by Dr. Joe Dispenza

THE BIOLOGY OF BELIEF 10TH ANNIVERSARY EDITION: Unleashing the Power of Consciousness, Matter & Miracles, by Bruce H. Lipton, Ph.D.

CRAZY SEXY KITCHEN: 150 Plant-Empowered Recipes to Ignite a Mouthwatering Revolution, by Kris Carr with Chef Chad Sarno

THE MINDBODY SELF: How Longevity Is Culturally Learned and the Causes of Health Are Inherited, by Dr. Mario Martinez

SOUL JOURNEYING: Shamanic Tools for Finding Your Destiny and Recovering Your Spirit, by Alberto Villoldo, Ph.D.

WHAT IF THIS IS HEAVEN?: How Our Cultural Myths Prevent Us from Experiencing Heaven on Earth, by Anita Moorjani

All of the above are available at your local bookstore,
or may be ordered by contacting Hay House (see next page).

We hope you enjoyed this Hay House book. If you'd like to receive our online catalog featuring additional information on Hay House books and products, or if you'd like to find out more about the Hay Foundation, please contact:

Hay House, Inc., P.O. Box 5100, Carlsbad, CA 92018-5100
(760) 431-7695 or (800) 654-5126
(760) 431-6948 (fax) or (800) 650-5115 (fax)
www.hayhouse.com® • www.hayfoundation.org

Published in Australia by:
Hay House Australia Pty. Ltd., 18/36 Ralph St., Alexandria NSW 2015
Phone: 612-9669-4299 • *Fax:* 612-9669-4144 • www.hayhouse.com.au

Published in the United Kingdom by:
Hay House UK, Ltd., Astley House, 33 Notting Hill Gate, London W11 3JQ
Phone: 44-20-3675-2450 • *Fax:* 44-20-3675-2451 • www.hayhouse.co.uk

Published in India by: Hay House Publishers India,
Muskaan Complex, Plot No. 3, B-2, Vasant Kunj, New Delhi 110 070
Phone: 91-11-4176-1620 • *Fax:* 91-11-4176-1630 • www.hayhouse.co.in

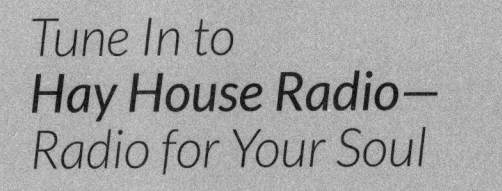